PROTEIN BIOCHEMISTRY, SYNTHESIS, STRUCTURE AND CELLULAR FUNCTIONS

METHIONINE

BIOSYNTHESIS, CHEMICAL STRUCTURE AND TOXICITY

PROTEIN BIOCHEMISTRY, SYNTHESIS, STRUCTURE AND CELLULAR FUNCTIONS

Additional books in this series can be found on Nova's website
under the Series tab.

Additional e-books in this series can be found on Nova's website
under the e-book tab.

MICROBIOLOGY RESEARCH ADVANCES

Additional books in this series can be found on Nova's website
under the Series tab.

Additional e-books in this series can be found on Nova's website
under the e-book tab.

PROTEIN BIOCHEMISTRY, SYNTHESIS, STRUCTURE AND CELLULAR FUNCTIONS

METHIONINE

BIOSYNTHESIS, CHEMICAL STRUCTURE AND TOXICITY

ALEXANDER SNEGURSKY
EDITOR

Nova Biomedical
New York

For permission to use material from this book please contact us:
Telephone 631-231-7269; Fax 631-231-8175
Web Site: http://www.novapublishers.com

Additional color graphics may be available in the e-book version of this book.

Library of Congress Cataloging-in-Publication Data

ISBN: 978-1-62618-246-2

Library of Congress Control Number: 2013933838

Published by Nova Science Publishers, Inc. † New York

Contents

Preface

Present monograph is dedicated to a very specific and interesting molecule/substance, i.e. to methionine. Methionine belongs to the α-amino acid group having chemical formula $HO_2CCH(NH_2)CH_2CH_2SCH_3$. (or, more concisely, – $C_5H_{11}NO_2S$). Being an essential amino acid, it is classified as nonpolar and one of the two sulfur-containing proteinogenic amino acids. Methionine is an intermediate in the biosynthesis of phospholipids, i.e. carnitine, cysteine, lecithin, phosphatidylcholine, taurine and others. It is also being utilized by plants in synthesizing ethylene. This amino acid is one of only two amino acids encoded by a single codon (AUG) in the standard genetic code. Therefore, methionine is incorporated into the N-terminal position of all proteins in eukaryotes and archaea during translation, although it is usually removed by post-translational modification. Being an essential amino acid, methionine is not synthesized in human organisms, while in plants and microorganisms it is synthesized with participance of aspartic acid and cysteine.

In this book, three major aspects of methionine are studied, namely, its biosynthesis, electronic structure and possible toxicity for live organisms. In this relation, the contributions of the present monograph cover all the above issues.

The synthesis of methione in plants has been studied recently in a number of works being summarized in a number of reviews. There are two alternative pathways leading to methionine synthesis in microorganisms: the transsulfuration pathway involves cystathionine as the intermediate and utilizes cysteine as the sulfur source, but the direct sulfhydrylation pathway bypasses cystathionine and uses inorganic sulfur instead. While most microorganisms synthesize methionine via either one of these pathways, Corynebacterium glutamicum utilizes both pathways, which appear to be fully functional. In C. glutamicum, each pathway is catalyzed by independent enzymes and is tightly regulated by methionine. Although the physiological significance of parallel pathways remains to be elucidated, their presence suggests metabolic flexibility and efficient adaptation of the organism to its environment.

The electronic structure of the methionine molecule has been studied in a number of original papers. It is a subject of increasing interest, partly because methods of theoretical calculation are reaching ever higher levels of sophistication and precision, and can predict the structure and function of complex molecules. Different structures of the methionine molecule have distinct physico-chemical properties. One of the most explicit properties of the structure of a molecule is its energy. The energy of a molecule is dependent on parameters such as its environment and conformation. Any calculation must be tested against experimental data to determine its reliability, and molecules in the gas phase represent a very suitable benchmark as they are not influenced by solid-state effects. However, the low vapor pressure may make it difficult to achieve sufficient sample densities to permit some spectroscopic measurements without thermally decomposing the compound. For this reason there is a lack of information about the electronic structure of methionine in the gas phase.

There are some evidences in literature related to the possible toxicity of methionine in human subjects. A number of nutritional and metabolic studies have employed amounts of methionine, including the *D* and *DL* isomers, both below and above the requirement and have not reported adverse effects in adults and children. Although methionine is known to exacerbate psychopathological symptoms in schizophrenic patients, there is no evidence of similar effects in healthy subjects. The role of methionine as a precursor of homocysteine is the most notable cause for concern. Longer-term studies in adults have indicated no adverse consequences of moderate fluctuations in dietary methionine intake, but intakes higher than 5 times normal resulted in elevated homocysteine levels. In infants, methionine intakes of 2–5 times normal resulted in impaired growth and extremely high plasma methionine levels, but no adverse long-term consequences were observed.

Present book is arranged in accordance with its title and scope. Nine chapters reflect almost equally the issues of the methionine biosynthesis, electronic structure and toxicity.

The first chapter by Biran and Ron deals with the physiological significance of methionine, the regulation of its synthesis and the evolution of methionine biosynthetic pathways. It stresses the fact that there are multiple enzymes for each step of the synthesis as well as a multitude of regulatory elements. This complexity probably reflects the importance of methionine in cellular metabolism and the necessity to monitor its cellular concentration with high sensitivity and precision.

The second chapter by Deminice et al.. is dedicated to some aspects of methione use for transmethylation reactions and homocysteine formation and removal. It also touches the problem of the methionine toxicity and the strategies to modulate methyl flux and homocysteine levels in several diseases.

In the third chapter by Boissonet et al., it has been shown that the developed model of iMSO-Fast and iMSO-Slow inbred mice enables one to study the mechanisms related to the appearance of MSO-dependent seizures allowing one to address the question how seizure sensitivity and brain glycogen content are related to each other. The proposed hypothesis suggests that brain glycogen metabolism may constitute a protection against epileptic attack, as glycogen may be degraded down to glucose-6-phosphate that can be used to either postpone the epileptic seizures or to provide neurons with energy when they need it. The ability of regulating this metabolism by variation of brain neurotransmitter contents, in particular the 5-HT neurotransmission, may constitute a defense against MSO-dependent epilepsy.

Amir and Hacahm emphasize in their chapter that methionine plays a significant role in plant metabolism. Through its metabolite, *S*-adenosyl-methionine (SAM), methionine is capable of controlling the levels of several key metabolites such as ethylene, polyamines and biotin. SAM is also the primary methyl group donor that regulates different processes in plants such as cell wall and chlorophyll synthesis. The level of methionine is tightly regulated in plants most probably because methionine associated metabolites control plant metabolism and growth.

The next section of the present monograph is dedicated to the electronic structure of methione. Three chapters deal with this problem.

Plekan et al. have reported the electronic structure of methionine being determined by means of photoemission and photoabsorption spectroscopy. It has been shown that the valence band spectroscopy yields the energies of the molecular orbitals responsible for the chemical bonding within the molecule. The methionine highest occupied molecular orbital (HOMO) has a sulfur lone pair character helping to stabilize the molecule against photofragmentation, compared with amino acids having aliphatic side chains. The core level spectra give information about the local chemical properties of each heavy atom in the molecule, while x-ray absorption sheds light on the unoccupied molecular orbitals. The two lowest orbitals accessible to carbon 1s excitation both have partial sulfur character. However neither of them is the lowest unoccupied molecular orbital (LUMO), which is instead the antibonding π orbital of the carboxylic acid group.

Next, the chapter by Myhovich et al. reports the photoluminescence data for methionine and absorption spectra of its aquatic solution measured using laser excitation at three different wavelengths –253, 337 and 380 nm – as well as applying the UV deuterium lamp. The main features of obtained luminescence have been traced. With the pumping laser wavelength increase the photoluminescence short-wave edge and the main maxima appeared to shift to the red region. For a number of luminescence spectra the Raman shift of pumping radiation was observed. The intensity of the Raman peaks gives new information about the relative intensity of the amino acid photoluminescence and the humidity of the amino acid powder. The absorption spectra give clear evidence that the properties of the methionine molecules in aquatic solutions are close to those in the solid state demonstrating the long-wave edge of absorption band. In addition, some energy characteristics of methionine have been obtained.

And finally, Snegursky et al. have studied both experimentally and theoretically the electron–impact induced fragmentation of the methionine molecule. The main emphasis has been given to the production of a number of ionized fragments from the initial molecule under study. A vast theoretical analysis allowed the main fragmentation channels of the methionine molecule to be elucidated and the basic geometric characteristics of this molecule structure to be found.

The last section of the present monograph is devoted to a very essential problem, i.e. the methonine toxicity. Kamata and Ishii have shown that methionine is the most toxic substance among the constituent amino acids of proteins. Excessive methionine intake causes oxidative/nitrosative injuries in the liver, hepatic/splenic hypertrophy, altered erythrocyte morphology, and the resultant splenic hemosiderosis in rats; however, the pathophysiology of methionine-induced liver toxicity remains to be elucidated. Methionine can be metabolized into cysteine by methionine cycle and transsulfuration, and cysteine can be further metabolized into cytoprotective metabolites such as glutathione, taurine, and hydrogen sulfide (H2S). It has been found that a methionine-rich diet induces acute lethal hepatitis in transsulfuration-defective cystathionine γ-lyase-deficient mice. Molecular mechanisms by which methionine excess induces hepatic dysfunction and physiological roles of transsulfuration in its detoxification (metabolism) are herein reviewed.

Dorszewska and Kozubski in the last chapter of this monograph demonstrate that antiepileptic drugs (AEDs) treatment in epileptics leads to increase in homocysteine and glutathione, decrease in cysteine and a bit unchanged methionine concentrations. Carbamazepine monotherapy, AEDs polytherapy and long-term therapy compared to control ones appear to be related to disturbances in both methionine: homocysteine and cysteine: homocysteine ratios. In patients with epilepsy before treatment with AEDs, the homocysteine and cysteine levels slightly change and significantly reduce the levels of methionine, and statistically insignificantly increase the levels of glutathione.

Summarizing, it should be noted that this monographs deals with the analysis of the most up-to-date data on the methionine molecule/substance properties related to its biosynthesis, electronic structure and toxicity issues. The editor believes that this book will be useful for students, post-graduates and scientists working in the filed of studying amino acids, being one the most prominent members of the family of molecules of biological relevance, tracing their basic properties and analyzing their practical applications in the field of pharmaceutics, medicine and molecular physics.

Alexander Snegursky
Editor

In: Methionine
Editor: Alexander Snegursky

ISBN: 978-1-62618-246-2

Chapter I

Methionine: A Key Regulatory Compound in Bacteria

***Dvora Biran*[1] *and Eliora Z. Ron*[*1,2]**
[1]Molecular Microbiology and Biotechnology, Tel Aviv University, Tel Aviv, Israel
[2]MIGAL-Galil Research Center, Israel

Abstract

In bacteria methionine is a key compound that regulates the synthesis of macromolecules and growth. The enzymes of methionine biosynthesis have a critical role in the adaptation to elevated temperatures and control the growth rate under these conditions. The genes coding for the biosynthetic methionine enzymes are under complex regulation which allows a quick response to changing physiological and environmental conditions. This chapter will discuss the physiological significance of methionine, the regulation of its synthesis and the evolution of methionine biosynthetic pathways. It will stress the fact that there are multiple enzymes for each step of the synthesis as well as a multitude of regulatory elements. This complexity probably reflects the importance of methionine in cellular metabolism and the necessity to monitor its cellular concentration with high sensitivity and precision.

Introduction

Methionine is a key amino acid – in addition to its role as a component of the proteins, it is the first amino acid in most proteins, in all kingdoms of life. It is therefore somewhat surprising that its biosynthesis was fully worked out relatively late and its regulation is still not fully understood. Moreover, as genetic-physiological studies of bacteria advance, additional regulatory circuits are discovered for methionine biosynthesis and activities. As of now there are evidence for the involvement of a repressor, an activator, feedback inhibition,

[*] Corresponding author: Eliora Z. Ron, E-mail: eliora@post.tau.ac.il

alternative heat shock sigma factor and RNA riboswitches. This complex regulation of methionine biosynthesis is probably essential for precise and sensitive regulation of the cellular concentration of this key compound.

Significance of Methionine in Bacteria

Methionine is a key compound - it is an amino acid that is present in most proteins and is the first amino acid in most proteins. Thus, the first codon in the majority of open reading frames codes for formyl-methionine or methionine and only in rare cases there is another first amino acid, usually leucine, isoleucine or valine. In the absence of methionine, protein synthesis stops and usually at initiation, thus preventing the accumulation of short truncated peptides. The cessation of protein synthesis leads to inhibition of RNA synthesis by stringent control and to inhibition of DNA synthesis. The latter is inhibited at initiation, as this step requires protein synthesis.

Methionine is the precursor of S-adenosyl methionine (SAM) which is the donor of C1 fragments for methylations and formylations. These C1 fragments are also used in the synthesis of pyrimidines, purines and fatty acids. The carbon chain of SAM is donated to putrescine, a precursor of the polyamines spermidine and spermine that neutralize the negative charge of nucleic acids. Recently another important physiological role was discovered for methionine – it is involved in the synthesis of autoinducer molecules such as N-acyl homoserine lactone, which are quorum-sensing signals and regulating gene expression in a cell density-dependent manner (for review see [1,2].

A scheme of the roles of methionine in bacteria is presented bellow.

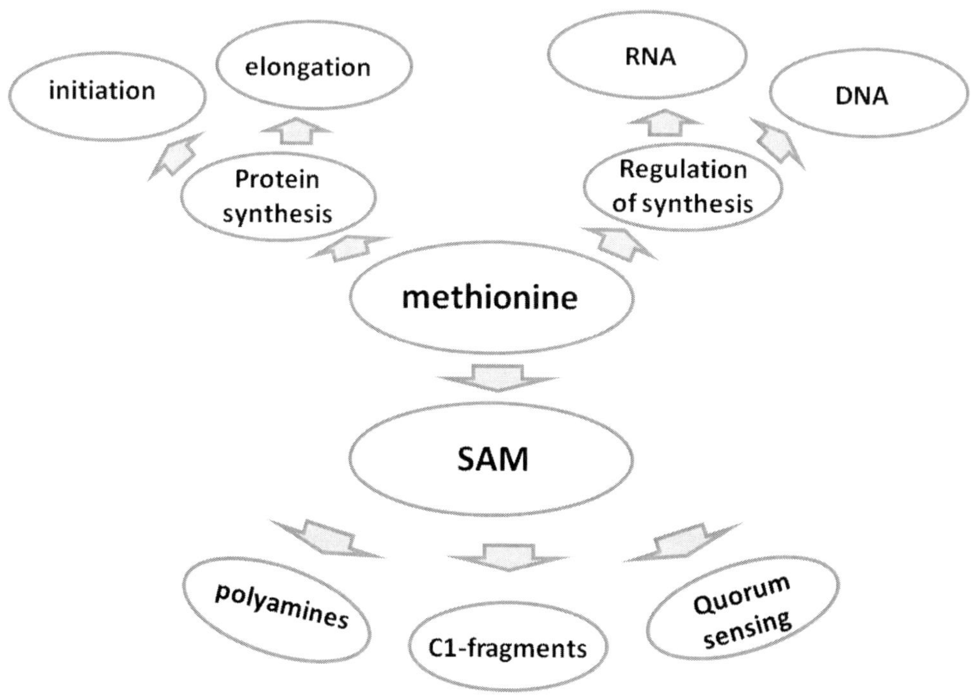

Methionine Biosynthesis

Methionine biosynthesis show a high degree of plasticity probably resulting from frequent events of duplication, functional reassignment, loss and regain throughout evolution. The end result is that each step in the biosynthesis is carried out by one of several possible enzymes. For review see [3,4]

The first step, homoserine activation, can occur in three ways - acetylation with acetyl-CoA by homoserine trans- acetylase, succinylation with succinyl-CoA, by homoserine trans-succinylase and phosphorylation with phosphate by homoserine kinase. So far, bacteria were shown to activate homoserine only by acetylation or succinylation. The enzymes which carry out the transacetylation are conserved and present in a large number of bacteria as well as fungi. The enzymes which carry out transsuccinylation are also conserved but are limited to fewer bacteria. Recent evidence indicated that enzymes of the homoserine transsucinylase family use acetyl-CoA and not succinyl-CoA if the glycine in position 111 is replaced by glutamate. Such a replacement occurs in *Bacillus cereus* and in *Agrobacterium tumefaciens* [5,6].

The second step of the pathway, incorporation of sulfur can also occur in three ways, by direct incorporation of sulfide into O-acetyl homoserine (sulfhydrylation) by direct incorporation of sulfide into O-succinyl homoserine (sulfhydrylation) and by transsulfuration, performed by cystathionine gamma-synthase with cysteine as sulfur donor, generating cystathionine. This is followed by beta elimination carried out by cystathionine beta-lyase forming homocysteine.

The final step is the methylation of homocysteine which can occur in two ways, defined by the enzyme: by cobalamin-dependent methionine synthase, coded for by the bacterial *metH* gene or by cobalamin-independent methionine synthase, coded for by the bacterial *metE* gene. In *Escherichia coli* there are two interesting findings about the two enzymes involved in this step – the product of the *metE* gene is one of the most abundant *E. coli* proteins [7] and metH is not induced upon methionine starvation [8]

To summarize, although the steps in methionine biosynthetic pathway are biochemically conserved, the enzymes involved do not form conserve sets. Thus, there is no single set of enzymes for methionine biosynthesis, but a "pool" of enzymes to pick from in a "mix and match" manner. In addition, for a given set of enzymes, sequence conservation is not sufficient to accurately predict enzymatic activities according to sequence similarity. For example, as already mentioned, homoserine trans-succinylases that are differ only in one amino acid, in position 111, will have different enzymatic activity, homoserine trans-succinylase or homoserine acetyl-transferase.

Regulatory Elements

Several control elements are involved in the regulation of methionine biosynthesis, for review see [8]. Similar to the methionine biosynthetic enzymes – one or more of these control elements is present in any given bacterial species.

Feedback inhibition of the first enzyme - in most bacteria the first enzyme in methionine biosynthesis (coding for homoserine trans-succinylase or for homoserine acetyl-transferase) is feedback-inhibited by methionine and its derivative SAM [9,10]. This is not always the case, as in Leptospira meyeri, the first enzyme is homoserine O-acetyltransferase, which is not feedback inhibited [11]. The feedback inhibition of the first biosynthetic enzyme has an immediate effect on methionine synthesis and its cellular level drops instantaneously.

MetJ is a transcriptional repressor which binds to specific sequences (Met Box) in the operator (-10 region) of methionine biosynthesis genes. This binding requires the co-repressor S-adenosylmethionine that promotes the binding of a MetJ dimer to the DNA [4,12,13].

SAM, MetK and GroESL– As SAM is directly involved both in the activity of the MetJ repressor and in the feedback inhibition of the first enzyme, its regulatory role is clear. MetK is methionine adenosyltransferase which catalyzes the formation of S-adenosylmethionine [14] and mutants deleted for *metK* are derepressed for methionine. A recent interesting finding indicates that *E. coli* MetK depends on the chaperonine GroE for folding [15,16]. Therefore, under conditions of GroE depletion the cellular level of MetK is reduced and the methionine pathway is induced [15].

MetR participates in the regulation of several genes involved in methionine biosynthesis [8]. It belongs to the LysR family of transcriptional regulators [17] and has a leucine zipper motif characteristic of many eukaryotic DNA-binding proteins in its N-terminal domain [18]. MetR is an autorepressor (represses the metR gene) and activates the *metE* and *metH* genes [8].

MetH is a cobalamin-dependent methionine synthase. Together with vitamin B12 it represses *metF* and *metE* [8]. It is active only under anaerobic conditions, as these are required for cobalamine (Vitamine B12) synthesis [19].

SAM riboswitch – Riboswitches are novel regulatory elements in the 5' leader sequences of bacterial transcripts regulate transcription of translation as a result of binding metabolites. Riboswitches that bind SAM and/or S-adenosylhomocysteine (SAH) were discovered in many bacteria [20-24] and so far there are eight distinct classes of such riboswitches that control various aspects of sulfur metabolism [25]. This finding stresses again the unusual diversity in the methionine biosynthetic pathway, as for most metabolites there is a single class of associated riboswitches [25].

The Complex Regulation of the First Enzyme in Methionine Synthesis

The extreme complexity of the regulatory systems for methionine biosynthesis is exemplified in the control of the synthesis, stability and activity of the first enzyme in the pathway of Gram negative bacteria.

In enteric bacteria the *metA* gene [26] codes for a homoserine trans-succinylase, and is feedback inhibited by methionine and SAM. At the transcriptional level it is repressed by MetJ and activated by MetR [27]. In addition, its transcription is also regulated by the heat shock transcription factor sigma-32 and is induced upon temperature upshifts [28]. In alpha-proteobacteria, *Agrobacterium tumefaciens*, there is an additional control at transcription, as *metA* has a SAM-riboswitch [23]. Thus, there are in Gram negative bacteria four different elements controlling *metA* synthesis at the transcriptional level.

The presence of a SAM riboswitch and a sigma-32 promoter on the same gene is interesting, as it offers the possibility to determine whether a riboswitch is effective in inhibiting an RNA-polymerase bound to sigma-32. Indeed, it was demonstrated that in *A. tumefaciens* methionine biosynthesis is also inhibited at 40°C, when transcription is driven with sigma32-RNA polymerase [6].

In addition to the multiple control elements involved in transcriptional regulation of *metA* the enzyme homoserine trans-succinylase (HTS) of *E. coli* is an unstable protein which undergoes regulated proteolysis by the cellular ATP-dependent proteases [29-33]. This finding is unexpected as most biosynthetic enzymes are stable, and further indicates the requirement for a tight regulation of methionine biosynthesis at all levels.

Maybe the most unusual property of MetA regulation involves its response to temperature. In *E. coli* this enzyme is highly thermo-labile and undergoes unfolding at temperatures above 20°C, with most unfolding occurring above 30°C. These temperatures are way lower than the unfolding temperatures of most – if not all - *E. coli* enzymes. Above 44°C the enzyme has very low activity and aggregates [34], resulting in methionine auxotrophy. Thus, the activity of *E. coli* HTS regulates the ability of these bacteria to grow as a function of elevated temperatures [35,36]. The regulation of growth at elevated temperature by the activity of HTS and availability of methionine is not limited to *E. coli* – it exists in all the enteric bacteria tested [31-33] and even extends to phylogenetically very distant bacteria, such as *Bacillus polymixa* [37,38].

Although the various bacterial strains vary in the response to temperature, the inhibition of growth at elevated temperature could be shown in all cases to be due to limitation in the endogenous supply of methionine. This uniformity of limiting growth rate in response to elevated temperature by limiting methionine synthesis could simply reflect an unusual genetic stability of the temperature-sensitive enzyme HTS. An alternative explanation for the temperature sensitivity of the methionine pathway in many bacterial species would assume that the ability to control the growth rate at elevated temperature can be advantageous to mesophilic bacteria and has been selected for independently in many bacterial strains.

Controlling growth rate by regulating the availability of methionine is efficient, as methionine is required for initiation and continuation of protein synthesis, biosynthesis of purines, various methylation reactions, and synthesis of polyamines. The evolutionary conservation of the regulation of growth rate by the availability of endogenous methionine is compatible with the idea that this regulatory system has been selected for.

Conclusions

Methionine is a key compound in bacteria that regulates the macromolecule synthesis and growth.

There are multiple enzymes for each step of methionine synthesis as well as a multitude of regulatory elements.

The multiple biosynthetic possibilities and the variety of control elements indicate the importance of methionine in cellular metabolism and the necessity to monitor its cellular concentration with high sensitivity and precision.

References

[1] Camilli A, Bassler BL (2006) Bacterial small-molecule signaling pathways. *Science* 311: 1113-1116.

[2] Venturi V (2006) Regulation of quorum sensing in Pseudomonas. *FEMS Microbiol Rev* 30: 274-291.

[3] Gophna U, Bapteste E, Doolittle WF, Biran D, Ron EZ (2005) Evolutionary plasticity of methionine biosynthesis. *Gene* 355: 48-57.

[4] Old IG, Phillips SE, Stockley PG, Saint Girons I (1991) Regulation of methionine biosynthesis in the Enterobacteriaceae. P*rog Biophys Mol Biol* 56: 145-185.

[5] Zubieta C, Arkus KA, Cahoon RE, Jez JM (2008) A single amino acid change is responsible for evolution of acyltransferase specificity in bacterial methionine biosynthesis. *J Biol Chem* 283: 7561-7567.

[6] Rotem O, Biran D, Ron EZ (2012) Methionine biosynthesis in *Agrobacterium tumefaciens*: study of the first enzyme. *Res Microbiol* in press.

[7] Tomoyasu T, Mogk A, Langen H, Goloubinoff P, Bukau B (2001) Genetic dissection of the roles of chaperones and proteases in protein folding and degradation in the *Escherichia coli* cytosol. *Mol Microbiol* 40: 397-413.

[8] Weissbach H, Brot N (1991) Regulation of methionine synthesis in *Escherichia coli* . *Mol Microbiol* 5: 1593-1597.

[9] Schlesinger S (1967) Inhibition of growth of *Escherichia coli* and of homoserine O-transsuccinylase by alpha-methylmethionine. *J Bacteriol* 94: 327-332.

[10] Lee LW, Ravel JM, Shive W (1966) Multimetabolite control of a biosynthetic pathway by sequential metabolites. *J Biol Chem* 241: 5479-5480.

[11] Bourhy P, Martel A, Margarita D, Saint Girons I, Belfaiza J (1997) Homoserine O-acetyltransferase, involved in the *Leptospira meyeri* methionine biosynthetic pathway, is not feedback inhibited. *J Bacteriol* 179: 4396-4398.

[12] He YY, McNally T, Manfield I, Navratil O, Old IG, et al.. (1992) Probing met repressor-operator recognition in solution. *Nature* 359: 431-433.

[13] Davidson BE, Saint Girons I (1989) The *Escherichia coli* regulatory protein MetJ binds to a tandemly repeated 8 bp palindrome. *Mol Microbiol* 3: 1639-1648.

[14] Markham GD, Hafner EW, Tabor CW, Tabor H (1980) S-Adenosylmethionine synthetase from *Escherichia coli* . *J Biol Chem* 255: 9082-9092.

[15] Fujiwara K, Taguchi H (2012) Mechanism of methionine synthase overexpression in chaperonin-depleted *Escherichia coli* . *Microbiology* 158: 917-924.

[16] Fujiwara K, Ishihama Y, Nakahigashi K, Soga T, Taguchi H (2010) A systematic survey of in vivo obligate chaperonin-dependent substrates. *EMBO J* 29: 1552-1564.

[17] Schell MA (1993) Molecular biology of the LysR family of transcriptional regulators. *Annu Rev Microbiol* 47: 597-626.

[18] Maxon ME, Wigboldus J, Brot N, Weissbach H (1990) Structure-function studies on *Escherichia coli* MetR protein, a putative prokaryotic leucine zipper protein. *Proc Natl Acad Sci U S A* 87: 7076-7079.

[19] Jeter RM, Olivera BM, Roth JR (1984) *Salmonella typhimurium* synthesizes cobalamin (vitamin B12) de novo under anaerobic growth conditions. *J Bacteriol* 159: 206-213.

[20] Heppell B, Blouin S, Dussault AM, Mulhbacher J, Ennifar E, et al.. (2011) Molecular insights into the ligand-controlled organization of the SAM-I riboswitch. *Nat Chem Biol* 7: 384-392.
[21] Fuchs RT, Grundy FJ, Henkin TM (2007) S-adenosylmethionine directly inhibits binding of 30S ribosomal subunits to the SMK box translational riboswitch RNA. *Proc Natl Acad Sci U S A* 104: 4876-4880.
[22] Fuchs RT, Grundy FJ, Henkin TM (2006) The S(MK) box is a new SAM-binding RNA for translational regulation of SAM synthetase. *Nat Struct Mol Biol* 13: 226-233.
[23] Corbino KA, Barrick JE, Lim J, Welz R, Tucker BJ, et al.. (2005) Evidence for a second class of S-adenosylmethionine riboswitches and other regulatory RNA motifs in alpha-proteobacteria. *Genome Biol* 6: R70.
[24] Winkler WC, Nahvi A, Sudarsan N, Barrick JE, Breaker RR (2003) An mRNA structure that controls gene expression by binding S-adenosylmethionine. *Nat Struct Biol* 10: 701-707.
[25] Batey RT (2011) Recognition of S-adenosylmethionine by riboswitches. *Wiley Interdiscip Rev RNA* 2: 299-311.
[26] Biran D, Michaeli S, Segal G, Ron EZ (1992) Location of the *metA* gene on the physical map of *Escherichia coli* . *J Bacteriol* 174: 5753-5754.
[27] Mares R, Urbanowski ML, Stauffer GV (1992) Regulation of the *Salmonella typhimurium metA* gene by the MetR protein and homocysteine. *J Bacteriol* 174: 390-397.
[28] Biran D, Brot N, Weissbach H, Ron EZ (1995) Heat shock-dependent transcriptional activation of the *metA* gene of *Escherichia coli* . *J Bacteriol* 177: 1374-1379.
[29] Gur E, Biran D, Ron EZ (2011) Regulated proteolysis in Gram-negative bacteria--how and when? *Nat Rev Microbiol* 9: 839-848.
[30] Biran D, Gur E, Gollan L, Ron EZ (2000) Control of methionine biosynthesis in *Escherichia coli* by proteolysis. *Mol Microbiol* 37: 1436-1443.
[31] Katz C, Rasouly A, Gur E, Shenhar Y, Biran D, et al.. (2009) Temperature-dependent proteolysis as a control element in *Escherichia coli* metabolism. *Res Microbiol* 160: 684-686.
[32] Ron EZ, Alajem S, Biran D, Grossman N (1990) *Adaptation of Escherichia coli to elevated temperatures: the metA gene product is a heat shock protein.* Antonie Van Leeuwenhoek 58: 169-174.
[33] Ron EZ (1975) Growth rate of Enterobacteriaceae at elevated temperatures: limitation by methionine. *J Bacteriol* 124: 243-246.
[34] Gur E, Biran D, Gazit E, Ron EZ (2002) In vivo aggregation of a single enzyme limits growth of *Escherichia coli* at elevated temperatures. *Mol Microbiol* 46: 1391-1397.
[35] Ron EZ, Shani M (1971) Growth rate of *Escherichia coli* at elevated temperatures: reversible inhibition of homoserine trans-succinylase. *J Bacteriol* 107: 397-400.
[36] Ron EZ, Davis BD (1971) Growth rate of *Escherichia coli* at elevated temperatures: limitation by methionine. *J Bacteriol* 107: 391-396.
[37] Wyman A, Shelton E, Paulus H (1975) Regulation of homoserine transacetylase in whole cells of *Bacillus polymyxa*. *J Biol Chem* 250: 3904-3908.
[38] Wyman A, Paulus H (1975) Purification and properties of homoserine transacetylase from *Bacillus polymyxa*. *J Biol Chem* 250: 3897-3903.

In: Methionine
Editor: Alexander Snegursky

ISBN: 978-1-62618-246-2

Chapter II

Methionine and Methylation Balance: Pathways for the Production and Removal of Homocysteine

Rafael Deminice* [*1,2], ***Flávia Troncon Rosa*** [2] ***and Alceu Afonso Jordao*** [2]

[1]Department of Physical Education, Faculty of Physical Education and Sport State University of Londrina, Londrina, Paraná, Brazil

[2]Laboratory of Nutrition and Metabolism, Faculty of Medicine of Ribeirao Preto, University of Sao Paulo, Ribeirao Preto, Sao Paulo, Brazil

Abstract

Methionine (Met) is an essential amino acid fundamentally involved in many biological processes. Met plays a pivotal role in methyl group metabolism and homeostasis through its conversion to S-adenosylmethionine (SAM). SAM acts primarily as a universal methyl donor in the synthesis of methylated compounds such as neurotransmitters (adrenaline, noradrenaline), phosphatidylcholine, L-carnitine, creatine and others. Among its many functions, SAM serves as the methyl group donor for all known mammalian DNA methyltransferases.Homocysteine (Hcy) is an amino acid formed exclusively by demethylation of Met. Tissue concentration of Hcy is maintained at normal levels by regulated production and efficient removal of this thiol. Specific enzymes coordinately regulate the flux through transmethylation and remethylation as well as transsulfuration pathways depending on the demand for SAM or Hcy accumulation. In the last few years, elevated concentration of Hcy in the blood has been considered a risk factor for the development of cardiovascular disease. Several studies have also related hyperhomocysteinemia (HHcy) to several diseases such as neurodegenerative diseases, diabetes, renal dysfunction, fatty liver disease and others. Recently, strategies for modulating the methylation flux and consequently decreasing or

* Corresponding Author: Rafael Deminice, Department of Physical Education, Faculty of Physical Education and Sport - State University of Londrina. Rodovia Celso Garcia Cid | Pr 445 Km 380 | Campus Universitário, Londrina, Paraná, Brazil. Phone: 05516 36024564, e-mail: deminice@ig.com.br

regulating Hcy levels have been studied. Creatine, betaine and choline are involved in the methyl donation pathway and have been shown to modulate SAM levels and the SAM/SAH ratio, as well as Hcy concentration in several conditions. Thus the aim of this review was to discuss Met utilization for transmethylation reactions and Hcy formation and removal. Arguing about Hcy toxicity and the recent strategies to modulate methyl flux and Hcy levels in several diseases and situations is also an objective of this review.

Key-words: Methionine, S-adenosylmethionine, homocysteine, methyl balance, thiol.

INTRODUCTION

Years ago, much of the interest in methionine metabolism has focused on homocysteine (Hcy) formation and its relation to cardiovascular and neurodegenerative diseases (Stipanuk, 2004). Over the last few years, several studies have also related hyperhomocysteinemia (HHcy) to diseases such as diabetes, renal dysfunction, fatty liver disease, bone health and others (Morris et al. 2005; Herrmann et al. 2005; Chen et al.. 2011; Wijekoon et al.. 2005; Taes et al.. 2004; Deminice et al.. 2011). Much of the explanation for this relation is based on the disturbance of methyl balance (Stead et al.. 2004; Brosnan et al.. 2007; ingenbleek, 2009) and toxicity of elevated Hcy levels (Weiss, 2005; Jakubowski, 2006). However, the exact mechanism and contribution of Hcy formation and excess to the above diseases are still poorly known.

Methionine is an essential amino acid, albeit a toxic one at high concentrations due to the increase and accumulation of Hcy (Jakubowski, 2006). Hcy formation is regulated by methionine requirements for transmethylation reactions through S-adenosylmethionine (SAM), a key intermediate in methionine metabolism (Stipanuk, 2004; Brosnan & Brosnan, 2006). Recently, strategies for modulating the methylation flux and consequently decreasing or regulating Hcy levels have been studied (Brosnan et al.. 2004; Stead et al.. 2006; Deminice et al.. 2009). B-vitamins, creatine, betaine and choline are involved in the methyl donation pathway and have been shown to modulate SAM levels and the SAM/S-adenosylhomocysteine (SAH) ratio as well as Hcy concentration in several conditions. Thus, the aim of this review was to discuss methionine utilization for transmethylation reactions and Hcy formation and removal. Arguing about Hcy toxicity and the recent strategies to modulate methyl flux and Hcy levels in several diseases and situations is also an objective of this review.

Methionine for Hcy Formation and Removal

Hcy is an amino acid of a complex nature, mainly occurring bound to albumin (Brosnan et al.., 2004). Hcy formation and removal involves three pathways transmethylation, remethylation and transsulfuration. Plasma and urinary Hcy values reflect the cellular synthesis, utilization and integrity of these metabolic pathways (Neves et al.., 2004). A summary of the three metabolic pathways of methionine is presented below.

Transmethylation and Homocysteine Formation

Hcy is formed exclusively from the demethylation of methionine from the diet or from catabolism of proteins. In Hcy synthesis, a considerable proportion of methionine is activated by ATP to form S-adenosylmethionine (SAM) in a reaction catalyzed by the enzyme methionine adenosyltransferase (MAT) (Brosnan et al.. 2004). Liver-specific (MAT 1 and MAT 3) and nonliver-specific (MAT 2) MAT exist. Expression of MAT 2 occurs in all tissues; however, the pancreas has the highest level of non-liver SAM formation (Stipanuk, 2004). SAM primarily acts as a universal methyl donor involving a variety of acceptors and is responsible for the biosynthesis of DNA, RNA, proteins, creatine, carnitine, phospholipids and others. S-adenosylhomocysteine (SAH), the subproduct of this methylation reaction, is hydrolyzed, generating adenosine and Hcy (Selhub, 1999). An alternative pathway of SAM is decarboxylation and donation of the aminopropyl groups for polyamine synthesis (spermine and spermidine) that consume less than 10% of SAM (Stipanuk, 2004). In addition, there is a considerable flux of SAM through glycine *N*-methyltransferase (GNMT). GNMT is important for the maintenance of SAM homeostasis, which utilizes glycine as the methyl receptor to consume excess SAM and forms the nontoxic product sarcosine (Martínez-Chantar et al.. 2008). Thus, in a condition of decreased SAM, GNMT may decrease the methylation of glycine to maintain SAM levels and the SAM/SAH ratio. Rowling et al.. (2002) showed that loss of GNMT in mice leads to the accumulation of hepatic SAM and to a marked increase in the hepatic SAM/SAH ratio. Conversely, activation of GNMT in rats by retinoic acid causes a reduction in SAM and global liver DNA methylation (Martínez-Chantar et al.. 2008).This sequence of reactions is referred to as transmethylation and is present in a variety of cells (Figure 1). This metabolism does not involve Hcy removal, which will be described next.

Hcy Removal by the Remethylation and Transsulfuration Pathways

Once formed, Hcy can be removed by two distinct pathways: remethylation and transsulfuration. The remethylation pathway allows methionine to be regenerated from Hcy using new methyl groups. This can occur via two enzymes: methionine synthase (MS) or betain-homocysteine methyltransferase (BHMT) that uses betaine as a substrate (Brosnan et al.., 2004). Reaction with tetrahydrofuran (THF) occurs in all tissues and utilizes folate as a substrate and vitamin B12 as a co-factor. $N^{5,10}$-methyl-THF is used as methyl donor though the action of the enzyme $N^{5,10}$-methyl-THF reductase (MTHFR). The methyl group used by MS is synthesized *de novo* in the folate coenzyme system using the 3-carbon of serine. This reaction is regulated by the bioavailability of SAM and SAH production, with SAM availability activating methyl donation from THF and SAH inhibiting it (Brosnan et al.. 2007).

Contrary to the MS and THF reaction, the donation of the methyl group of betaine via BHMT does not depend on vitamin B12 and folate (Williams & Sclalinske, 2007). BHMT is only present in the liver and kidneys of mammals and requires dietary betaine or betaine derived from choline as the methyl donor (Stipanuk, 2004). BHMT has an important role in the overall conversion of Hcy to methionine in the remethylation pathway, although it is considered by some authors as the alternative pathway for Hcy remethylation (Stipanuk, 2004; Brosnan et al.. 2007). This is because MS responds by donating most of the methyl groups for Hcy remethylation back to methionine in the liver (Martinov et al.. 2000). However, BHMT has an impact on the methyl balance and plays a regulatory role in Hcy homeostasis (Willians & Schalinske, 2007). Collinsova et al.. (2006) found that in vivo inhibition of BHMT in mice causes reduction of liver SAM and HHcy. Studies have shown increased BHMT mRNA abundance in models of impaired folate-dependent remethylation (Townsend et al.. 2004; Shinohara et al.. 2010).

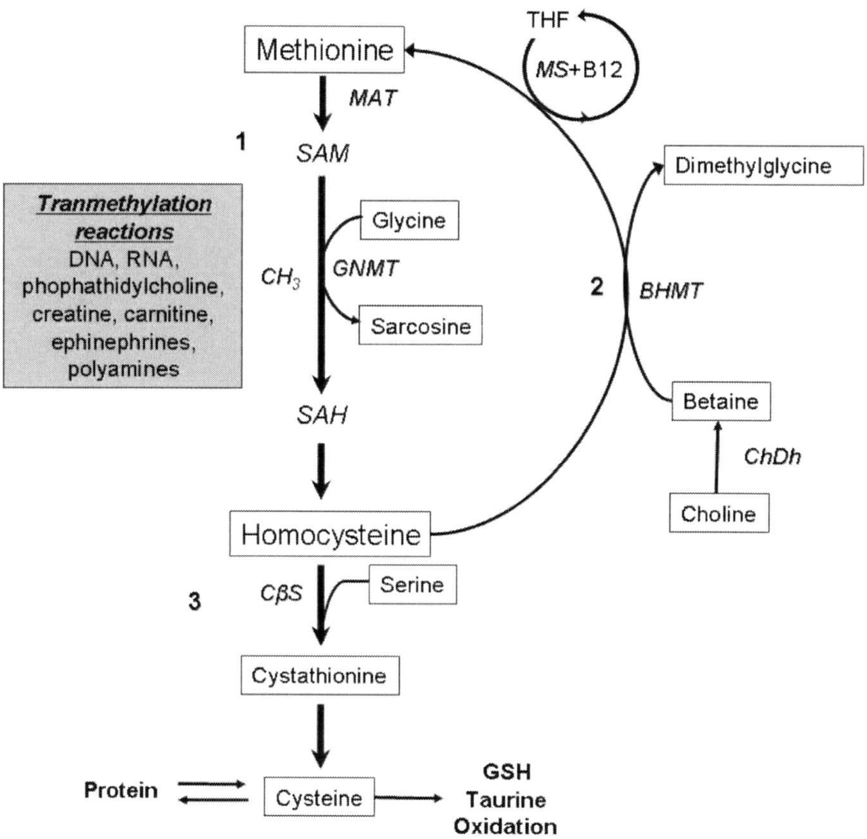

Figure 1. Methionine metabolism and Hcy formation. 1 transmethylation; 2 remethylation and 3 transsulfuration. SAM = S-adenosylmethionine; SAH = S-adenosylhomocysteine; MS = methionine synthase, GNMT = glycine N-methyltransferase, BHMT = betaine-homocysteine methyltransferase, CβS = cystathionine β-synthase, ChDh = choline dehydrogenase, MAT = methionine adenosyltransferase.

The cycle described above does not involve the catabolism of methionine, which occurs by transsulfuration. In this process, Hcy reacts with serine to form cystathione in an irreversible reaction catalyzed by the enzyme cystathione-β-synthase (CβS) and then hydrolized to form cysteine. Thus, the transsulfuration pathway is responsible for both Hcy catabolism and cysteine formation. The transsulfuration process is limited to the liver, kidneys, small bowel and pancreas (Brosnan et al.. 2004). Excess cysteine is oxidized by taurine or excreted into the urine (Brosnan et al.. 2007). Transsulfuration is also important for the synthesis of reduced glutathione (GSH), an important endogenous antioxidant (Feoli et al.., 2006). Thus, an important link is observed between methyl balance and cell redox state (Prudova et al.. 2006). Studies have shown that down-regulation in the transsulfuration pathway may increase Hcy concentration and decrease GSH production, affecting the redox status of the cells and resulting in lipid peroxidation and oxidative stress (Mosharov et al.. 2000; Prudova et al.. 2006; Feoli et al.. 2008).

Methyl Balance and Regulation of Methionine Metabolism

Methionine metabolism is regulated by two major mechanisms: 1) allosteric regulation of SAM and 2) Hcy concentration and regulation of expression of key enzymes (Brosnan & Brosnan, 2006). Martinov et al.. (2000), in a mathematical model, stated that when SAM levels are low, priority for this substrate goes to transmethylation reactions, promoting reduction of the transsulfuration pathway. Also, changes in the rate of methyl consumption lead to correspondingchanges in the rate of SAMproduction (Finkelstein, 2007). In contrast, high levels of SAM, in a high protein or high methionine diet for example, promote feedback of repression of MAT and SAM production. Thus, in this instance, SAMinhibits its own production.In addition,SAM is also an activator of CβS and an inhibitor of MS (Brosnan & Brosnan, 2006). Therefore, excess methionine may increase the availability of SAM and Hcy. This situation promotes activation of transsulfuration and inhibition of remethylation pathways (Stipanuk, 2004), whereas approximately70% of Hcy is catabolized by the enzyme CβS, in contrast to what occurs with protein-free diets or during fasting, when the remethylation pathway is favored and only 10% Hcy is catabolized by CβS in an attempt to maintain the SAM pool (Deminice et al.. 2009).

GNMT is also an important regulator of transmethylation because it consumes excess SAM and maintains SAM homeostasis, which utilizes glycine as the methyl receptor and formssarcosine (Martínez-Chantar et al.. 2008). Thus, in a condition of increased SAM levels, GNMT may increase the methylation of glycine to sarcosine to decrease SAM levels and the SAM/SAH ratio (Finkelstein, 2007). Studies in GNMT knockout mice have shown impaired methionine metabolism characterized by increased methionine and SAM levels that caused fat liver accumulation and hepatocellular carcinoma (Martínez-Chantar et al.. 2008; Varela-Rey et al.. 2010).

In addition to allosteric regulation of SAM, the flux of methionine to remethylation versus transsulfuration is regulated by the concentration of Hcy and the expression of its key enzymes (CβS, BHMT and MS). Changes in CβS, BHMT and GNMT activities were found in liver of rats fed high methionine diets (Kim et al.. 2003). The *Km* of CβS for Hcy tends to be higher than those of the enzymes of remethylation (Stipanuk, 2004). Therefore, high Hcy concentrations activate transsulfuration at the expense of remethylation. Conversely, low Hcy formation increases the flux through the remethylation pathway (Selhub et al.., 2000). In addition, the transsulfuration pathway appears to be sensitive to the redox state of the cells (Vitvitsky et al.. 2003). As described above, in addition to catabolizing the excess of Hcy, the transsulfuration pathway also provides cysteine and glutathione (Stipanuk, 2004) and cysteine availability is a limiting factor of glutathione synthesis (Prudova et al.. 2006; Mosharov et al.. 2000). Thus, the balance between antioxidants and prooxidants also regulates GSH synthesis and the flux of Hcy to the transsulfuration pathway. Increased peroxides lead to increases in transsulfuration flux, whereas decreased oxidants reduce it (Mosharov et al.. 2000).

In remethylation, BHMT works as a complementary system to the folate-dependent system. MS donates most of the methyl groups to form methionine from Hcy remethylation (Nieman et al.. 2006). Besides being complementary, BHMT is important for overall remethylation especially when the folate-dependent system is impaired. BHMT activity is increased in MTHFR (–/–) knockout mice (Schawahn et al.. 2003) and micropigs fed folate-deficient diets (Halsted et al.. 2002), showing the complementary activity of BHMT to remethylation in folate depletion situations.Dietary betaine supplements induce BHMT activity, preventing severe hepatic steatosis in null MTHFR mice, while reducing Hcy levels by about 40% (Parajes &Pérez-Sala, 2006).

Hyperhomocysteinemia and Homocysteine Tocixity

Since McCully (1969) first described the relation between hyper-homocysteinemia (HHcy), atherosclerosis and arterial and venous thrombosis in patients with homocysteinuria, cumulative evidence has been emerged regarding Hcy as an independent risk factor for cardiovascular diseases, dementia, loss of cognitive functions and renal dysfunctions in patients with end-stage disease (Wyss & Schulze, 2002; Persky & Brazeau, 2001; Taes et al.., 2004). A relation between high Hcy concentrations in blood and several diseases such as diabetes, renal dysfunction, fatty liver disease, bone conditions, cancer and others has been recently demonstrated (Morris et al. 2005; Herrmann et al. 2005; Chen et al.. 2011; Wijekoon et al.. 2005; Taes et al.. 2004; Deminice te al. 2011; Mohamed et al.. 2011). It has also been suggested that altered Hcy levels may promote oxidative injury (Hogg, 1999; Huang, 2001; Streck et al.., 2003), affect the antioxidant defense system and promote lipid peroxidation (Blundel et al.., 1996).

The production of free radicals and reactive oxygen species (ROS) may contribute to a variety of diseases or may be present in several conditions of toxicity (Halliwel, 1997). Studies using experimental models of HHcy and endothelial cells incubated with Hcy have indicated that high levels of this amino acid may promote ROS formation, especially the superoxide anion (O_2^-) (Young et al.., 1997; Huang et al.., 2001; Streck et al.., 2003; Weiss, 2005) by auto-oxidation of Hcy and/or cysteine (Hogg, 1999) and that this increase in the intracellular free radical production may cause cell injuries (Streck et al.., 2003). It has been reported that Hcy may also cause important disorders in the system of antioxidant defense (Blundell et al.., 1996) regarding the enzymes glutathione peroxidase (GPx) and superoxide dismutase (SOD), mechanisms that may contribute to the increased induction of oxidative stress (Weiss, 2005) (Figure 2). In addition, the elevated ROS production may react with nitric oxide (NO) from the vascular endothelium, forming peroxynitrite ($OONO^-$). This reaction results in a reduction of NO bioactivity, inducing dysfunctions of vasodilation and of antiaterothrombosis mechanisms (Weiss, 2005).

Chemically, Hcy is a non-protein amino acid that contains a free sulfhydryl group (R-SH) known as "thiol". Cysteine is the most abundant thiol in plasma. Under aerobic conditions (in the presence of molecular oxygen as an electron receptor) and at physiological pH, Hcy and cysteine are oxidized in the form of disulfide according to the following general reaction:

$$2\,RSH + O_2 \rightleftharpoons RSSR + H_2O_2 \qquad (1)$$

This reaction is accompanied by the generation of ROS such as hydrogen peroxide (H_2O_2) and superoxide anion (O_2^-) (Weiss, 2005).Hcy, through its auto-oxidation by the mechanism described above, can oxidize other "thiols" such as cysteine and GSH to form mixed disulfides. Thus, an altered relation among all the aminothiols in plasma has been demonstrated in experimental models of HHcy (Mansoor et al.., 1993). However, the potential of Hcy for promoting ROS formation *in vivo*is little known.

Hogg (1999) stated that auto-oxidation of excess Hcy represents a mechanism by which aminothiols participate in ROS formation and consequently in pathological events. It has been demonstrated that HHcy increases thiobarbituric acid reactive substances (TBARS) and reduces the antioxidant capacity of tissues (TRAP) *in vitro*. The author also suggested that changes induced by HHcy in these biomarkers are a strong indication of oxidative stress and it results from an imbalance between free radical production and antioxidant defenses (Jacob et al.., 1998; Huang et al.., 2001; Heydrick et al.., 2004). Thus, cumulative data indicate that HHcy can promote oxidative stress, cause important disorders in the antioxidant system and trigger proinflammatory processes.

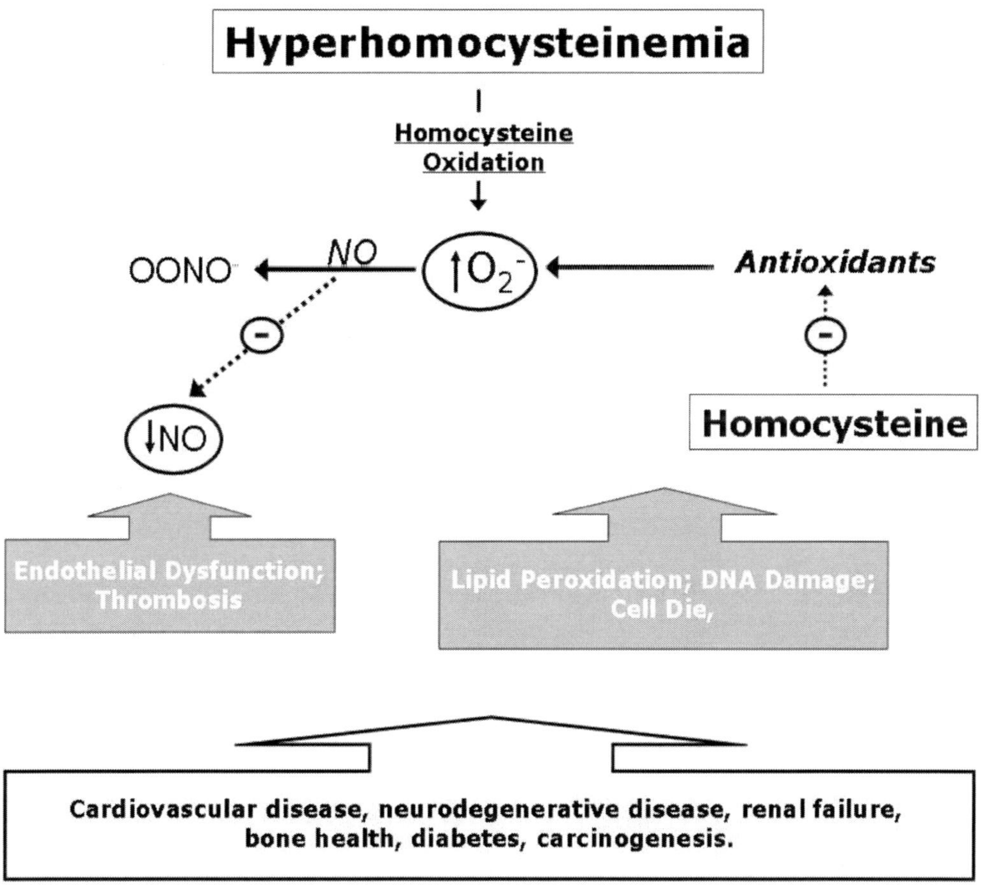

Figure 2. Hypothetic mechanism of the generation of reactive oxygen species and pathological events by homocysteine oxidation.

Modulation of the Methylation Demand and Hcy Production

Although there are many methyltransferases that utilize SAM (probably 50 to100 methyltranferases known) (Stead et al.. 2006), about 5 are the major ones responsible for Hcy regulation and accumulation (Willians & Schalinske, 2007). Remethylation by folate-dependent (MS and MTHFR) and non-folate dependent (BHMT) systems has a great impact on methyl balance and Hcy formation. Also,guanidinoacetate N-methyltransferase (GAMT) and phosphatidylethanolamine N-methyltransferase (PEMT) are considered to be the largest consumers of SAM in the formation of creatine and phosphatidylcholine, respectively.Therefore, b-vitamin, betaine, choline and creatine deprivation or or supplementation has been shown to modulate the methylation flux and consequently Hcy formation. Also, some authors have demonstrated the relation between methyl balance modulation, Hcy formation and several diseases (Deminice et al.. 2011; Chiu et al.. 2009; Pizzolo et al.. 2011; Bazzano et al.. 2002; Ishihara et al.. 2008).

Altered Methyl Balance by B-Vitamin Deficiency and Supplementation

Folate and vitamin B12 deficiencies are probably the most studied factors related to methionine metabolism and the supplementation of these vitamins is a common strategy for the prevention of hyperhomocysteinemia (Debreceni et al.. 2012). In remethylation, $N^{5,10}$-methyl-THF is used as a methyl donor though the action of the enzyme MTHFR. MTHFR synthesizes a major methyl donor for Hcy remethylation to methionine metabolism using folate as a substrate and vitamin B12 as a cofactor (Selhub et al.. 2000; Stipanuk et al.. 2004). The methyl group used by MS is synthesized *de novo* in the folate coenzyme system using the 3-carbon of serine. This reaction is regulated by the bioavailability of SAM and SAH production, with SAM availability activating methyl donation from THF and SAH inhibiting it. Deficiencies in folate or vitamin B12 or deletion in enzymes of the folate cycle (MS and MTHFR) have been related to several disturbances such as deficit in fatty acid oxidation and fatty liver (Pooya et al.. 2012), neurological disturbances (Prasad et al.. 2011; Jadavji et al.. 2012), cardiomyopathies (Garcia et al.. 2011), multiple myeloma (Kim et al.. 2007) and others. Impaired removal of Hcy by the transsulfuration pathway is also related to several disturbances (Jiang et al.. 2012; Zhang et al.. 2009; Hamelet et al.. 2007). Decreased activity or inborn errors in CβS are among the most common causes of high Hcy levels in blood (Brosnan et al.. 2007). Vitamin B6 deficiency can also cause hyperhomocysteinemia and related diseases because the action of CβS depends on this vitamin (Stipanuk, 2004).

In contrast, several studies have shown that folate and vitamin B supplementation could diminish or prevent disturbances of methionine metabolism and HHcy caused by vitamin deficiency or deletion in enzymes involved is methionine metabolism (Peeters et al.. 2007; Pizzolo et al.. 2011; Debreceni et al.. 2012). B-vitamins and folate supplementation have been shown to decrease Hcy in animal models of HHcy (Peeters et al.. 2007; Kolling et al.. 2011, Moens et al.. 2008; Qipshidze et al.. 2010) and in several conditions affecting humans such as hemodialysis (Chiu et al.. 2009), hyperhomocysteinemia related to polymorphism of MTHFR677TT(Pizzolo et al.. 2011), cardiovascular disease (Bazzano et al.. 2002; Ishihara et al.. 2008), and neurological dysfunctions (Morris et al.. 2006; Malouf et al.. 2008). Studies on MTHFR mutation have provided evidence of the association between HHcy and cardiovascular disease (Selhub et al.. 2000). B-vitamin complex supplementation has been effective in decreasing and controlling Hcy levels in individuals with MTHFR, MS and CβSmutation. It has also been tested for their cardioprotective role (Selhub et al.. 2000; Pizzolo et al.. 2011). Recent studies have shown the cardioprotective effects of folic acid in rats (Moens et al.. 2008) and mice (Qipshidze et al.. 2010).

The folic acid fortification policy has opened a new dimension in the prevention of vascular diseases (Debreceni et al.. 2012). In the UnitedStates and Canada, folic acid fortification of enriched grain products was fully implemented in 1998. Yang et al.. evaluated the trends in stroke-related mortality comparing the United States and Canada to England and Wales where fortification is not required. These authors concluded that the improvement in strokemortality observed after folic acid fortification in the UnitedStates and Canada but not in England and Wales is consistent with the hypothesis that folic acid fortification reduces deaths from stroke. However, recent reviews and meta-analyses have discussed the ineffective action of Hcy-lowering B-vitamin in the prevention of cardiovascular diseases (Carlsson et al.. 2006; Debreceni et al.. 2012; Huang et al.. 2012). These authors concluded that the results are uncertain in justifying the effectiveness of vitamin substitution in the prevention of cardiovascular and cerebrovascular diseases. Huang et al.. (2012) stated that B vitamin supplementation can exert a significant protective effect on stroke but is not demonstrably effective in the secondary prevention of cardiovascular disease or in the improvement of all cause mortality.

Methyl Balance Modulation by the Methyl Donors Choline and Betaine

Choline and betaine are metabolically connected to both lipid and folate-dependent one-carbon metabolism (Ueland, 2011). Choline and betaine are obtained from the diet or by de novo synthesis in tissues and most of the produced choline (~95%) is used for phosphatidylcholine (PC) synthesis. In liver, PC is synthesized through two routes: 1) the CDP-choline pathway, for which the substrate is dietary choline, and 2) methylation of phosphatidylethanolamine catalyzed by PEMT, which requires 3 molecules of SAM (Li & Vance, 2008). In the liver and kidney, choline is oxidized to betaine through the action of two enzymes, i.e., choline dehydrogenase (ChDh) and betaine aldehyde dehydrogenase (BaDh) (Pajaresa & Pérez-Salab, 2006). The formation of betaine links choline to one-carbon metabolism, because betaine serves as a methyl donor in the BHMT reaction (Figure 3) (Ueland, 2011). Thus, betaine promotes Hcy remethylation to methionine and thereby affects the concentration of the universal methyl donor SAM and choline is responsible for phosphatidylcholine synthesis and SAM consumption via PEMT.

Studies have shown that a choline-deficient diet (CDD) impairs methionine metabolism by decreasing SAM availability and the SAM/SAH ratio by reducing BHMT and GNMT gene expression and consequently increasing Hcy levels (Setoue et al.. 2008). CDD also increases liver fat content and liver injury (Raubenheimer et al.. 2006; Kulinski et al.. 2004). Choline is required for the synthesis of PC, an essential part of very low density lipoproteins (VLDL). Several studies have shown that CDD impairs the synthesis of phosphatidylcholine and reduces the secretion of liver triglyceride such as VLDL, resulting in accumulation of liver triglycerides (Li & Vance, 2008;Macfarlane et al.. 2010; Oishi et al.. 2011).Deminice et al.. (2012) observed that CDD promoted accumulation of total fat, triglycerides and cholesterol in the liver. Also, increased liver malondialdehyde (MDA) and plasma alanine aminotransferase (ALT) were found in CDD-fed rats compared to control rats. These changes were accompanied by increased Hcy and decreased liver SAM and SAM/SAH ratio as well as smaller BHMT and GNMT mRNA abundance in CDD-fed rats. The use of CDD and of a choline-methionine-deficient diet (CMDD) is considered to be an important study model for non-alcoholic steatohepatitis (NASH). CDD also decreases betaine availability and BHMT activity (Raubenheimer et al.. 2006). Collinsova et al.. (2006) found that *in vivo* inhibition of BHMT in mice causes reduction of liver SAM and HHcy. Teng et al.. (2011) were the first to report a successful induction of BHMT deletion in mice and detected decreased SAM and choline and increased Hcy and GNMT and ChDh activities. Theses authors also concluded that BHMT deletion may promote a disturbance of methionine metabolism, with the occurrence of HHcy, fatty liver and hepatocellular carcinoma. Thus, there is supporting evidence showing that choline and/or betaine depletion causes remethylation (via BHMT) and PC formation disturbances with several consequences such as HHcy, steatohepatitis and hepatocellular carcinoma.

In contrast, it is known that supplementation with betaine and/or choline decreases Hcy (Steenge et al.. 2003) and protects the liver from fat accumulation and lipid peroxidation in rodent models of alcoholic (Erma et al.. 2004; Kim et al.. 2008) and non-alcoholic (Kwon et al.. 2009) steatosis. Previous studies have shown that betaine administration to mice and rats increased MAT activity, hepatic methionine and SAM as well as decreased Hcy concentrations (Kim et al.. 2003; Kim et al.. 2005). Kwon et al.. (2009) demonstrated that betaine supplementation protects the liver from nonalcoholic steatosis and oxidative stress most probably via its effects on SAM availability and transsulfuration reactions in rats fed a high-fat diet for 3 weeks. Betaine supplementation also regulates phosphatidylcholine synthesis and normalizes VLDL production rates, preventing fat accumulation in the liver induced by ethanol ingestion (Kharbanda et al.. 2009). Dietary betaine or supplements also raise BHMT activity, preventing severe hepatic steatosis in null MTHFR mice, while reducing Hcy levels by about 40% (Parajes & Pérez-Sala, 2006). Atkinson et al.. (2008) observed that dietary and supplementary betaine and choline acutely increased plasma betaine and attenuated the post-methionine load rise in Hcy concentrations in humans. The effect of betaine supplementation on Hcy concentration has been tested also in clinical settings. In general, betaine supplementation (6-20 g per day) reduces fasting and post-methionine plasma Hcy levels (Wilcken et al.. 1983; Berlow et al..1984; Wendel et al.. 1989; Abdelmalek et al.. 2009). Thus, most of the protective effects of betaine have been attributed to the fact that it is an important methyl donor via BHMT, resulting in the conversion of Hcy to methionine with a subsequent increase in hepatic SAM synthesis in several situations.

Synthesis and Supplementation of Creatine: A methyl balance modulator

Creatine is a methylated compound and creatine synthesis is responsible for a considerable SAM consumption in the liver (Brosnan et al.. 2011).The first step in creatine synthesis involves the reversible transfer of the amidino group of arginine to glycine to form guanidinoacetic acid (GAA) and ornithine in a reaction catalyzed by the enzyme arginine:glycine amidinotransferase (AGAT) which is very active in the kidneys. Next, the irreversible transfer of a methyl group from SAMto GAA is catalyzed by the enzyme GAMT (Brosnan and Brosnan 2007). The products of this reaction are creatine and SAH. The metabolism of creatine is connected to that of Hcy (Taes et al.., 2003). Creatine synthesis is responsible for a considerable consumption of SAM in the liver and Hcy formation (Stead et al.. 2006). Mudd & Pole (1975) and Mudd et al.. (1980) have calculated that this process consumes more SAM than all of the other methyltranferases combined (~75%). However, over the last few years, several authors have pointed out controversies in the classical studies conducted by Mudd & Pole (1975) and Mudd et al.. (1980) and have suggested that creatine synthesis may not be the main factor responsible for the methyltransferase reactions. Recent studies have indicated that phosphatidyl synthesis by the enzyme PEMT, together with creatine synthesis, accounts for the main consumption of SAM (Noga et al.. 2003; Stead et al.. 2004; Stead et al.., 2006). Noga et al.. (2003), while investigating the role of PEMT in Hcy metabolism using knockout mice, detected a significant fall in plasma Hcy compared to control and concluded that a significant portion of SAM is utilized in the synthesis of phosphatidylcholine. Stead et al.. (2006), after pointing out some gaps in the classical studies of Mudd and Poole (1975) and Mudd et al.. (1980)and based on studies using stable isotopes to determine the methylation flux, concluded that the synthesis of phosphatidylcholine is the major consumer of SAM. These authors also recalculated the methylation balance and pointed out that the synthesis of creatine is responsible for approximately 40% of the methylation demand, i.e., the estimated rate of creatine synthesis was significantly reduced compared to the pioneering studies about the methylation balance in humans (Mudd and Poole, 1975; Mudd et al.. 1980). Mudd et al.. (2007) stated that more evidence is needed to establish the relative contribution to the methylation flux by the synthesis of phosphatidylcholine, creatine and sarcosine. In addition, recent studies have demonstrated that other important substances may modulate the methylation balance and the levels of Hcy, such as betaine, glycine and serine (Fukada et al.., 2006). Thus, today we know that phosphatidylcholine synthesis and creatine synthesis are quantitatively the most prominent SAM consumers in the liver (Li & Vance, 2008). Phospholipid methylation is thought to be the major SAM-consuming pathway since hepatic PC synthesis by PEMT is responsible for approximately 40% of hepatic Hcy formation (Li & Vance, 2008, Jacobs et al.. 2008). Creatine synthesis is also responsible for a very considerable consumption of hepatic SAM (Stead et al.. 2006), perhaps as much as 40%.

Based on the above considerations, it has been demonstrated that creatine supplementation modulates methylation demand (Edison et al.. 2007) and decreases plasma Hcy concentration in rats (Edison et al.. 2007; Deminice et al.. 2009). Creatine supplementation increases plasma, muscle and liver creatine concentration. As a consequence, a remarkable (~80%) down-regulation of renal AGAT activity and decreased GAA activity are found, showing a reduction in the endogenous formation of creatine (Staed et al.. 2001; Edison et al.. 2007; Deminice et al.. 2011). Since creatine synthesis is responsible for a considerable consumption of methyl groups and Hcy formation (Stead et al.. 2006), creatine synthesis suppression by creatine supplementation leads to decreasing plasma Hcy in different situations (Stead et al.. 2001; Taes et al.. 2003; Edison et al.. 2007; Deminice et al.. 2009; Deminice et al.. 2011). Stead et al.. (2001) studied how methylation demand and Hcy formation were modulated by creatine and GAA intake using rats receiving either 0.4% creatine, 0.36% GAA or a control diet. Hcy levels were reduced by 25% by creatine, and increased by 50% by GAA, while plasma Met levels were unchanged. Taes et al.. (2003) also detected significant reductions in plasma Hcy levels in a group of nephrectomised rats fed 2% creatine for two weeks, as well as a significant negative correlation between Hcy and plasma creatine. Also, creatine supplementation decreased both Hcy and lipid peroxidation in rats (Deminice et al.. 2009). Lately, Deminice et al.. (2011) reported that four weeks of creatine supplementation (2% in the diet) inhibited increased Hcy levels after acute exercise in rats.

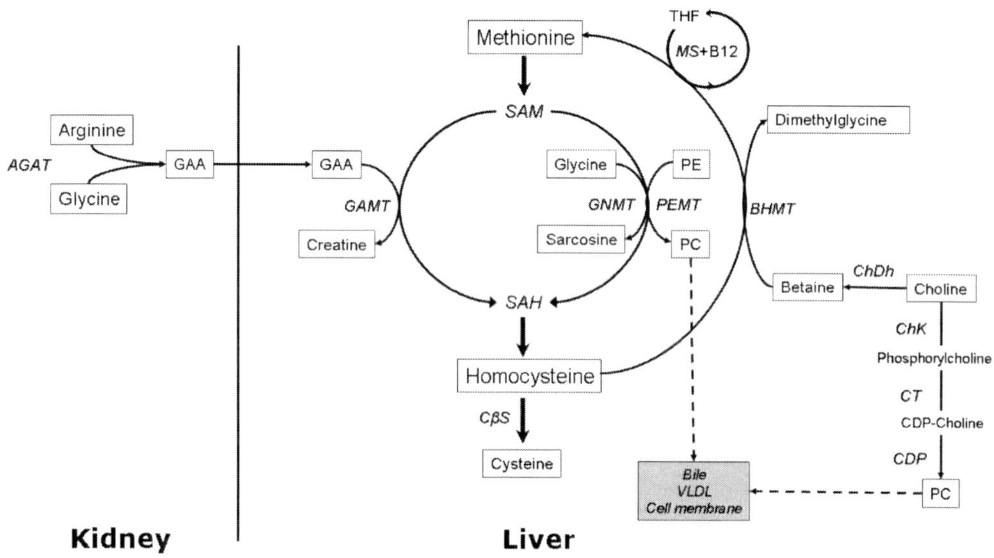

Figure 3.Creatine, phosphatidylcholine and choline metabolic interactions with methionine metabolism and Hcy formation. AGAT, arginine:glycine amidinotransferase; GAMT, S-adenosylmethionine:guanidinoacetate *N*-methyltransferase; GAA, guanidinocetate; SAM, S-adenosylmethionine; SAH, S-adenosylhomocysteine; PEMT, phosphatidylethanolamine *N*-methyltransferase; PE, phosphatidlyethanolamine; PC, phosphatidylcholine; CT, CTP:phosphorylcholinecytidylyltransferase; BHMT, betaine-homocysteine S-methyltransferase.

More recently, data from our laboratory showed that creatine supplementation normalized mRNA of GNMT and BHMT that were disturbed by two different models of steatosis in rats (Deminice et al.. 2011; Deminice et al.. 2012). A high-fat and choline-deficient diet leads to fatty liver, decreased SAM and SAM/SAH ratio, increased Hcy and lipid peroxidation. Creatine supplementation prevented these perturbations by methyl balance modulation as demonstrated by the restoration of GNMT and BHMT mRNA abundance.

Summary

In summary, disturbances or imbalance of methionine metabolism are important metabolic events regarding health. Hcy is the key intermediary in methionine metabolism and has been considered as a marker of disturbance and diseases. Recently, b-vitamins, betaine, choline and creatine deprivation or supplementation have been shown to modulate the methylation flux and consequently Hcy formation. Also, some authors have demonstrated the relation between methyl balance modulation, Hcy formation and several diseases including fatty liver disease, diabetes and hepatocellular carcinoma. Thus, a clear understanding of how we can modulate the methyl balance and Hcy formation is essential in order to minimize adverse effects and optimize health.

References

Abdelmalek MF, Sanderson SO, Angulo P, Soldevila-Pico C, Liu C, Peter J, Keach J, Cave M, Chen T, McClain CJ, Lindor KD. Betaine for nonalcoholic fatty liver disease: results of a randomized placebo-controlled trial.*Hepatology*. 2009 Dec;50(6):1818-26.

Banerjee R, Zou CG. Redoxregulation and reactionmechanism of human cystathionine-beta-synthase: a PLP-dependent hemesensor protein.*Arch Biochem Biophys*. 2005 Jan 1;433(1):144-56.

Bazzano LA, He J, Ogden LG, Loria C, Vupputuri S, Myers L, Whelton PK. Dietary intake of folate and risk of stroke in US men and women: NHANES I Epidemiologic Follow-up Study. National Health and Nutrition Examination Survey. *Stroke* 2002;33:1183–1188.

Berlow S, Bachmann RP, Berry GT, et al.. Betaine therapy in homocystinuria. *Brain Dysfunction*. 1989;2:10–24.

Brosnan JT, Brosnan ME. The sulfur-containing amino acids: an overview.*J Nutr*. 2006 Jun;136(6 Suppl):1636S-1640S.

Brosnan JT, Jacobs RL, Stead LM, Brosnan ME. Methylation demand: a key determinant of homocysteine metabolism.*Acta Biochim Pol*.2004;51(2):405-13.

Brosnan JT, da Silva R, Brosnan ME. Amino acids and the regulation of methyl balance in humans.*Curr Opin Clin Nutr Metab Care*. 2007 Jan;10(1):52-7.

Chiu YW, Chang JM, Hwang SJ, Tsai JC, Chen HC. Pharmacological dose of vitamin B12 is as effective as low-dose folinic acid in correcting hyperhomocysteinemia of hemodialysis patients.*Ren Fail*. 2009;31(4):278-83.

Carlsson CM. Homocysteine lowering with folic acid and vitamin B supplements: effects on cardiovascular disease in older adults. *Drugs Aging*. 2006;23(6):491-502.

Debreceni B, Debreceni L. Why do homocysteine-lowering B vitamin and antioxidant e vitamin supplementations appear to be ineffective in the prevention of cardiovascular diseases?*Cardiovasc Ther*. 2012 Aug;30(4):227-33.

Deminice R, da Silva RP, Lamarre SG, Brown C, Furey GN, McCarter SA, Jordao AA, Kelly KB, King-Jones K, Jacobs RL, Brosnan ME, Brosnan JT. Creatine supplementation prevents the accumulation of fat in the livers of rats fed a high-fat diet.*J Nutr*. 2011 Oct;141(10):1799-804.

Deminice R, Vannucchi H, Simões-Ambrosio LM, Jordao AA. Creatine supplementation reduces increased homocysteine concentration induced by acute exercise in rats.*Eur J Appl Physiol*. 2011 Nov;111(11):2663-70.

Deminice R, Portari GV, Vannucchi H, Jordao AA. Effects of creatine supplementation on homocysteine levels and lipid peroxidation in rats.*Br J Nutr*. 2009 Jul;102(1):110-6.

Deminice R, Francisco LV, da Silva LECM, Frajacomo FT, Garcia SB, Jordao AA. Creatine supplementationand homocysteine metabolismo in rats fed choline-deficient diet. *Clin Chem Lab Med*. 2012;50(12):A37.

Erman F, Balkan J, Cevikbaş U, Koçak-Toker N, Uysal M. Betaine or taurine administration prevents fibrosis and lipid peroxidation induced by rat liver by ethanol plus carbon tetrachloride intoxication.Amino Acids 27: 199-205, 2004.

Kim SJ, Jung YS, Kwon do Y, Kim YC. Alleviation of acute ethanol-induced liver injury and impaired metabolomics of S-containing substances by betaine supplementation.Biochem Biophys Res Commun 368: 893-8, 2008.

Finkelstein JD. Metabolic regulatory properties of S-adenosylmethionine and S-adenosylhomocysteine.*Clin Chem Lab Med.* 2007;45(12):1694-9.

Garcia MM, Guéant-Rodriguez RM, Pooya S, Brachet P, Alberto JM, Jeannesson E, Maskali F, Gueguen N, Marie PY, Lacolley P, Herrmann M, Juillière Y, Malthiery Y, Guéant JL. Methyl donor deficiency induces cardiomyopathy through altered methylation/acetylation of PGC-1α by PRMT1 and SIRT1.*J Pathol.* 2011 Nov;225(3):324-35.

Halsted CH, Villanueva JA, Devlin AM, Niemelä O, Parkkila S, Garrow TA, Wallock LM, Shigenaga MK, Melnyk S, James SJ. Folate deficiency disturbs hepatic methionine metabolism and promotes liver injury in the ethanol-fed micropig.*Proc Natl Acad Sci U S A.* 2002 Jul 23;99(15):10072-7

Hamelet J, Demuth K, Paul JL, Delabar JM, Janel N. Hyperhomocysteinemia due to cystathionine beta synthase deficiency induces dysregulation of genes involved in hepatic lipid homeostasis in mice.*J Hepatol.* 2007 Jan;46(1):151-9.

Herrmann M, Widmann T, Herrmann W. Homocysteine--a newly recognised risk factor for osteoporosis.*Clin Chem Lab Med.* 2005;43(10):1111-7.

Huang T, Chen Y, Yang B, Yang J, Wahlqvist ML, Li D. Meta-analysis of Bvitaminsupplementation on plasmahomocysteine, cardiovascular and all-cause mortality.*Clin Nutr.* 2012 Aug;31(4):448-54.

Ingenbleek Y. Hyperhomocysteinemia is a biomarker of sulfur0deficiency in human morbidities. *Open Clin Chemest J*; 2:49-60, 2009

Ishihara J, Iso H, Inoue M, et al.. Intake of folate, vitamin B6 and vitamin B12 and the risk of CHD: The Japan Public Health Center-Based Prospective Study Cohort I. *J Am Coll Nutr* 2008;27:127–136.

Jadavji NM, Deng L, Leclerc D, Malysheva O, Bedell BJ, Caudill MA, Rozen R.. Severe methylenetetrahydrofolate reductase deficiency in mice results in behavioral anomalies with morphological and biochemical changes in hippocampus.*Mol Genet Metab.* 2012 Jun;106(2):149-59.

Jiang H, Stabler SP, Allen RH, Maclean KN. Altered expression of apoA-I, apoA-IV and PON-1 activity in CBS deficient homocystinuria in the presence and absence of treatment: Possible implications for cardiovascular outcomes.*Mol Genet Metab.* 2012 May 5. [Epub ahead of print].

Jacobs,RL,Stead LM, Devlin C, Tabas I, Brosnan ME, Brosnan JT, Vance D. E.(2005)Physiological regulation of phospholipid methylation alters plasma homocysteine in mice. *J. Biol. Chem.* 280, 28299–28305.

Kim YG, Kim SK, Kwon JW, Park OJ, Kim SG, Kim YC, Lee MG. Effects of cysteine on amino acid concentrations and transsulfuration enzyme activities in rat liver with protein-calorie malnutrition.*Life Sci.*2003 Jan 24;72(10):1171-81.

Kim HN, Kim YK, Lee IK, Lee JJ, Yang DH, Park KS, Choi JS, Park MR, Jo DY, Kim HJ. rphisms involved in the folate metabolizing pathway and risk of multiple myeloma.*Am J Hematol.* 2007 Sep;82(9):798-801.

Kolling J, Scherer EB, da Cunha AA, da Cunha MJ, Wyse AT. Homocysteine induces oxidative-nitrative stress in heart of rats: prevention by folic acid. *Cardiovasc Toxicol.* 2011 Mar;11(1):67-73.

Kulinski A, Vance DE, Vance JE. A choline-deficient diet in mice inhibits neither the CDP-choline pathway for phosphatidylcholine synthesis in hepatocytes nor apolipoprotein B secretion.*J Biol Chem.* 2004 Jun 4;279(23):23916-24.

Lawson-Yuen A, Levy HL. The use of betaine in the treatment of elevated homocysteine.*Mol Genet Metab.* 2006 Jul;88(3):201-7.

Macfarlane DP, Zou X, Andrew R, Morton NM, Livingstone DE, Aucott RL, Nyirenda MJ, Iredale JP, Walker BR. Metabolic pathways promoting intrahepatic fatty acid accumulation in methionine and choline deficiency: implications for the pathogenesis of steatohepatitis.*Am J Physiol Endocrinol Metab.* 2011 Feb;300(2):E402-9.

Malouf R, Grimley Evans J. Folic acid with or without vitamin B12 for the prevention and treatment of healthy elderly and demented people.*Cochrane Database Syst Rev.* 2008 Oct 8;(4):CD004514.

Martínez-Chantar ML, Vázquez-Chantada M, Ariz U, Martínez N, Varela M, Luka Z, Capdevila A, Rodríguez J, Aransay AM, Matthiesen R, Yang H, Calvisi DF, Esteller M, Fraga M, Lu SC, Wagner C, Mato JM. Loss of the glycine N-methyltransferase gene leads to steatosis and hepatocellular carcinoma in mice.*Hepatology.* 2008 Apr;47(4):1191-9.

Martinov MV, Vitvitsky VM, Mosharov EV, Banerjee R, Ataullakhanov FI. A substrate switch: a new mode of regulation in the methionine metabolic pathway.*J Theor Biol.* 2000 Jun 21;204(4):521-32.

McNeil CJ, Hay SM, Rucklidge GJ, Reid M, Duncan G, Maloney CA, Rees WD. Disruption of lipid metabolism in the liver of the pregnant rat fed folate-deficient and methyl donor-deficient diets.*Br J Nutr.* 2008 Feb;99(2):262-71.

Mosharov E, Cranford MR, Banerjee R. The quantitatively important relationship between homocysteine metabolism and glutathione synthesis by the transsulfuration pathway and its regulation by redox changes.*Biochemistry.* 2000 Oct 24;39(42):13005-11.

Morris MS, Jacques PF, Selhub J. Relation between homocysteine and B-vitamin status indicators and bone mineral density in older Americans.*Bone.* 2005 Aug;37(2):234-42.

Morris MC, Evans DA, Schneider JA, Tangney CC, Bienias JL, Aggarwal NT. Dietary folate and vitamins B-12 and B-6 not associated with incident Alzheimer's disease.*J Alzheimers Dis.* 2006 Aug;9(4):435-43.

Moens AL, Champion HC, Claeys MJ, Tavazzi B, Kaminski PM, Wolin MS, Borgonjon DJ, Van Nassauw L, Haile A, Zviman M, Bedja D, Wuyts FL, Elsaesser RS, Cos P, Gabrielson KL, Lazzarino G, Paolocci N, Timmermans JP, Vrints CJ, Kass DA. High-dose folic acid pretreatment blunts cardiac dysfunction during ischemia coupled to maintenance of high-energy phosphates and reduces postreperfusion injury.*Circulation.* 2008 Apr 8;117(14):1810-9.

Mudd SH, Brosnan JT, Brosnan ME, Jacobs RL, Stabler SP, Allen RH, Vance DE, Wagner C. Methyl balance and transmethylation fluxes in humans.*Am J Clin Nutr.* 2007 Jan;85(1):19-25.

Nieman KM, Hartz CS, Szegedi SS, Garrow TA, Sparks JD, Schalinske KL. Folate status modulates the induction of hepatic glycine N-methyltransferase and homocysteine metabolism in diabetic rats.*Am J Physiol Endocrinol Metab.* 2006 Dec;291(6):E1235-42.

Oishi T, Terai S, Iwamoto T, Takami T, Yamamoto N, Sakaida I. Splenectomy reduces fibrosis and preneoplastic lesions with increased triglycerides and essential fatty acids in rat liver cirrhosis induced by a choline-deficient L-amino acid-defined diet.*Hepatol Res.* 2011 May;41(5):463-74.

Pajares MA, Pérez-Sala D.Betaine homocysteine S-methyltransferase: just a regulator of homocysteinemetabolism?*Cell Mol Life Sci.* 2006 Dec;63(23):2792-803.

Peeters AC, van Aken BE, Blom HJ, Reitsma PH, den Heijer M. The effect of homocysteine reduction by B-vitamin supplementation on inflammatory markers. *Clin Chem Lab Med.* 2007;45(1):54-8.

Pizzolo F, Blom HJ, Choi SW, Girelli D, Guarini P, Martinelli N, Stanzial AM, Corrocher R, Olivieri O, Friso S. Folic acid effects on s-adenosylmethionine, s-adenosylhomocysteine, and DNA methylation in patients with intermediate hyperhomocysteinemia. *J Am Coll Nutr.* 2011 Feb;30(1):11-8.

Pooya S, Blaise S, Moreno Garcia M, Giudicelli J, Alberto JM, Guéant-Rodriguez RM, Jeannesson E, Gueguen N, Bressenot A, Nicolas B, Malthiery Y, Daval JL, Peyrin-Biroulet L, Bronowicki JP, Guéant JL. Methyl donor deficiency impairs fatty acid oxidation through PGC-1α hypomethylation and decreased ER-α, ERR-α, and HNF-4α in the rat liver. *J Hepatol.* 2012 Apr 17. [Epub ahead of print].

Prasad AN, Rupar CA, Prasad C. Methylenetetrahydrofolate reductase (MTHFR) deficiency and infantile epilepsy.*Brain Dev.* 2011 Oct;33(9):758-69.

Qipshidze N, Tyagi N, Sen U, Givvimani S, Metreveli N, Lominadze D, Tyagi SC. Folic acid mitigated cardiac dysfunction by normalizing the levels of tissue inhibitor of metalloproteinase and homocysteine-metabolizing enzymes postmyocardial infarction in mice.*Am J Physiol Heart Circ Physiol.* 2010 Nov;299(5):H1484-93.

Rowling MJ, McMullen MH, Schalinske KL. Vitamin A and its derivatives induce hepatic glycine N-methyltransferase and hypomethylation of DNA in rats. *J Nutr* 2002; 132:365-369

Raubenheimer PJ, Nyirenda MJ, Walker BR. A choline-deficient diet exacerbates fatty liver but attenuates insulin resistance and glucose intolerance in mice fed a high-fat diet.*Diabetes.* 2006 Jul;55(7):2015-20.

Schwahn BC, Wendel U, Lussier-Cacan S, Mar MH, Zeisel SH, Leclerc D, Castro C, Garrow T, Rozen R. Effects of betaine in a murine model of mild cystathionine-beta-synthase deficiency. *Metabolism* 2004: 53, 594–599.

Schwahn, B. C., Chen, Z., Laryea, M. D., Wendel, U., Lussier-Cacan, S., Genest, J. Jr, Mar, M. H., Zeisel, S. H., Castro, C., Garrow, T. and Rozen, R. (2003) Homocysteine-betaine interactions in a murine model of 5,10-methylenetetrahydrofolate reductase deficiency. *FASEB J.* 17, 512–514.

Selhub J, Jacques PF, Bostom AG, Wilson PW, Rosenberg IH. Relationship between plasma homocysteine and vitamin status in the Framingham study population. Impact of folic acid fortification.*Public Health Rev.* 2000;28(1-4):117-45.

Shinohara M, Ji C, Kaplowitz N. Differences in betaine-homocysteine methyltransferase expression, endoplasmic reticulum stress response, and liver injury between alcohol-fed mice and rats. *Hepatology.* 2010 Mar;51(3):796-805.

Stead LM, Jacobs RL, Brosnan ME, Brosnan JT. Methylation demand and homocysteine metabolism.*Adv Enzyme Regul.* 2004;44:321-33.

Stead LM, Brosnan JT, Brosnan ME, Vance DE, Jacobs RL. Is it time to reevaluate methyl balance in humans?*Am J Clin Nutr*. 2006 Jan;83(1):5-10.

Steenge GR, Verhoef P, Katan MB. Betaine supplementation lowers plasma homocysteine in healthy men and women.*J Nutr*. 2003 May;133(5):1291-5

Stipanuk MH. Sulfur amino acid metabolism: pathways for production and removal of homocysteine and cysteine.*Annu Rev Nutr*. 2004;24:539-77.

Teng YW, Mehedint MG, Garrow TA, Zeisel SH. Deletion of betaine-homocysteine S-methyltransferase in mice perturbs choline and 1-carbon metabolism, resulting in fatty liver and hepatocellular carcinomas.*J Biol Chem*. 2011 Oct 21;286(42):36258-67.

Townsend JH, Davis SR, Mackey AD, Gregory JF. Folate deprivation reduces homocysteine remethylation in a human intestinal epithelial cell culture model: role of serine in one-carbon donation.*Am J Physiol Gastrointest Liver Physiol*.2004 Apr;286(4):G588-95.

Ueland PM. Choline and betainein health and disease.*J Inherit Metab Dis*. 2011 Feb;34(1):3-15.

Varela-Rey M, Martínez-López N, Fernández-Ramos D, Embade N, Calvisi DF, Woodhoo A, Rodríguez J, Fraga MF, Julve J, Rodríguez-Millán E, Frades I, Torres L, Luka Z, Wagner C, Esteller M, Lu SC, Martínez-Chantar ML, Mato JM. Fatty liver and fibrosis in glycine N-methyltransferase knockout mice is prevented by nicotinamide.*Hepatology*. 2010 Jul;52(1):105-14.

Williams KT, Schalinske KL. New insights into the regulation of methyl group and homocysteine metabolism.*J Nutr*. 2007 Feb;137(2):311-4.

Wilcken DE, Wilcken B, Dudman NP, Tyrrell PA. Homocystinuria--the effects of betaine in the treatment of patients not responsive to pyridoxine.*N Engl J Med*. 1983 Aug 25;309(8):448-53.

Wendel U, Bremer HJ. Betaine in the treatment of homocystinuria due to 5, 10-methylenetetrahydrofolate reductase deficiency. *Eur J Pediatr*.1984;142:147–150.

Yang Q, Botto LD, Erickson JD, Berry RJ, Sambell C, Johansen H, Friedman JM. Improvement in stroke mortality in Canada and the United States, 1990 to 2002.*Circulation*. 2006 Mar 14;113(10):1335-43.

Zhang D, Jiang X, Fang P, Yan Y, Song J, Gupta S, Schafer AI, Durante W, Kruger WD, Yang X, Wang H. Hyperhomocysteinemia promotes inflammatory monocyte generation and accelerates atherosclerosis in transgenic cystathionine beta-synthase-deficient mice.*Circulation*. 2009 Nov 10;120(19):1893-902.

In: Methionine
Editor: Alexander Snegursky

ISBN: 978-1-62618-246-2

Chapter III

Selected Fast and Slow Methionine Sulfoximine-Inbred Mice: MSO-Dependent Seizures, Behavior, Brain Glycoprofiles, Neurotransmitters and Glycogen

Arnaud Boissonnet[1,3], Tobias Hévor[1], Richard Daniellou[2], Ludovic Landemarre[3] and Jean-François Cloix[1,2]

[1]Laboratoire de Neurobiologie, Université d'Orléans, Rue de Chartres, Orléans CEDEX 2, France

[2]Institut de Chimie Organique et Analytique, Université d'Orléans, Rue de Chartres, Orléans CEDEX 2, France

[3]GLYcoDiag, Chevilly, France

Abstract

Methionine sulfoximine (MSO) is a byproduct due to oxidation of methionine during bleaching of wheat used for animal nutrition. The compound was reported to alter animal behavior. Among these alterations, MSO has been described as epileptogenic for a variety of animals. In addition, MSO was glycogenic affecting especially the brain. In order to study the putative relationships between these two effects, we 1) analyzed the concomitant effect of MSO related to seizure induction and brain glycogen content in various strains of mice, and 2) developed strains of mice based upon the latency towards MSO-dependent generalized convulsions. Mice with short latency period were designated as MSO-Fast strain whereas those exhibiting a long latency period were designated as MSO-Slow strain. These two strains which had been obtained after an extensive selection process including inbreeding (giving either iMSO-Fast or iMSO-Slow mice) were used for the study of MSO induced responses in terms of alteration in seizure latency, neurotransmitter concentration, glycogen content, behavior, and in terms of the cerebral glycol-profile in cortical astrocyte cell membrane proteins.

In terms of sensitivity, MSO-Fast mice exhibited a very short latency toward MSO-dependent seizures when challenged with 75 mg/kg MSO (MSO-Fast n=20, 277.6 ± 64.5 min) as compared to MSO-Slow ones which did not responded to that dose during the first 10 hours after application. The EEG confirmed these data. Kainic acid was the only other convulsant that demonstrated a significant difference between the two strains of mice, MSO-Fast being significantly more reactive than MSO-Slow ones (MSO-Fast: 42 ± 30 min; MSO-Slow: 75 ± 52 min, $p<0.05$, n=20). In MSO-Fast mice MSO-induced seizures were significantly antagonized by MK-801. Therefore, we proposed that the glutamatergic pathway is involved in MSO-dependent seizures in our mouse model.

Behavioral open field testing revealed that iMSO-Fast mice were more anxious than iMSO-Slow strain. These observed differences might be related to the differences detected for the cerebral serotonin (5-HT) contents among both strains since brain of iMSO-Fast mice contained 1.48 ± 0.65 pmole/mg protein while that of iMSO-Slow ones contained 0.59 ± 0.31 pmole/mg protein, $p<0.01$, (n=8) as revealed by reversed-phase HPLC coupled to electro-chemical detection,

As observed for 5-HT the basal glycogen content was significantly higher in iMSO-Fast mice (8.4 ± 1.5 nmoles glucosyl/mg protein) compared to iMSO-Slow ones (3.2 ± 1.0 nmoles glucosyl/mg protein, $p<0.001$, n=10). The time course during which glycogen increased after MSO administration was significantly elevated in MSO-Fast mice than in MSO-Slow ones, while no significant variations were detected in both strains during control kinetics. These differences are supported by our *in vitro* studies addressing glycosylation by glycoprofiling with lectins on astrocytes.

Our results showed that the developed model of iMSO-Fast and iMSO-Slow inbred mice enabled us to study the mechanisms related to the appearance of MSO-dependent seizures allowing us also to address the question how seizure sensitivity and brain glycogen content are related to each other. Our current hypothesis suggests that brain glycogen metabolism may constitute a protection against epileptic attack, as glycogen may be degraded down to glucose-6-phosphate that can be used to either postpone the epileptic seizures or to provide neurons with energy when they need it. The ability of regulating this metabolism by variation of brain neurotransmitter contents, in particular the 5-HT neurotransmission, may constitute a defense against MSO-dependent epilepsy.

Keywords: Methionine sulfoximine, epilepsy, seizure, neurotransmitter, glycogen, mouse, glycosylation

I. Introduction

Epilepsies express as a variety of symptoms, such as absences, partial and generalized seizures. The World Health Organization (WHO) recognizes at least 40 forms of epilepsies (Dreifuss, 1987, (CCTILAE), 1989, Everitt and Sander, 1999, Wieser et al.., 2001). The symptoms correspond to a sudden and temporary synchronization of neuronal activity, whose origins are not well understood (Johnston, 1996, Dudek et al.., 1999, McCormick and Contreras, 2001, Rampp and Stefan, 2006, Uhlhaas and Singer, 2006, Scharfman, 2007). Medical imagery of epilepsies is based upon various mechanisms; the uptake of glucose analogs representing one of them (Engel et al.., 1982, Engel et al.., 1983, Sitoh and Tien, 1998, Lai et al.., 2010). This allowed the demonstration that epileptic foci are hyper-metabolic during the ictal period and hypometabolic during the inter-ictal period (Engel et al.., 1982, Engel et al.., 1983, Lai et al.., 2010). Such a hypometabolism of epileptic foci might be due to an intrinsic increase in the capacity for brain energy production as a consequence of various alterations in the cell such as glycolysis, gluconeogenesis or glycogenesis, the latter resulting in an accumulation of glycogen. That increase in glycogen content is difficult to demonstrate in humans. Nevertheless, brain glycogen content could now be considered as an index of brain activity, and is regarded as a diagnostic and therapeutic tool. In the future this could be useful when the inherent difficulties of measuring brain glycogen content using non-invasive methodologies, both in humans and animals are overcome (Gruetter, 2002, Gruetter et al.., 2003, Oz et al.., 2003, Brown, 2004, Brown et al.., 2005, Oz et al.., 2007, Cloix et al.., 2008, van Heeswijk et al.., 2010). However, it has been reported that brain biopsies obtained from the hippocampus of epileptic patients contained high glycogen content compared to grey and white matter from the same individuals (Dalsgaard et al.., 2007). During seizures, when neurons need a lot of energy, since they are starved by their high activities, this energy could come from different sources: blood supply, neuronal glycolysis and/or from a glial source especially from astrocytes. When the first two are outpaced, as astrocytes are the major cells capable of glycogen synthesis and accumulation, the astrocytic glycogen could be mobilized, and astrocyte glycolysis of the G-6-P derived from glycogen could then generate lactate, which is transferred to neurons. Neurons are therefore provided with a metabolite that can be used directly by the mitochondria for ATP production without further processing (Cloix et al.., 2008, Cloix and Hevor, 2009).

Among the various epileptic or seizure-induced models, the MSO-model is one which is chemically induced and in which both associated seizures and a specific increase in brain glycogen content is observed (Folbergrova et al.., 1969, Folbergrova, 1973, Hevor and Delorme, 1991, Hevor, 1994, Bernard-Hélary et al.., 2000, Bernard-Hélary et al.., 2002, Cloix and Hevor, 2009, Cloix et al.., 2010a). This model corresponds to seizures induced by methionine sulfoximine (MSO).

I.1. Historical Background

During the 50's, white food was used for feeding animals instead of crude food. The process used for producing the so called white food was a process designated as "bleaching" or "agenizing" which referred to the oxidation of grain products by the use of nitrogen trichloride. Such a process leads to the degradation of proteins up to the release of amino-acids; among them methionine. The latter of which became oxidized to MSO. Later MSO had been identified as the chemical responsible for the behavior abnormalities observed in animals fed with white food. Among abnormalities seizures were reported and described as the most relevant pathology (Gastaut et al.., 1958, Wolfe and Elliot, 1962, Wada et al.., 1967).

I.2. MSO-Induced Seizures

Animals, especially rats and mice, developed seizures after intraperitoneal administration of a single convulsive dose of MSO, and such seizures resembled the human "grand mal", the most striking and debilitating form of human epilepsy. In treated animals epileptiform seizures developed during a period of 6-8 hours after MSO dosing (pre-convulsive period), followed by recurrent seizures during 24-48 hours (convulsive or ictal period or *status epilepticus*) and then they recovered normal behavior (post-convulsive period) (Hevor, 1994, Bernard-Hélary et al.., 2000, Bernard-Hélary et al.., 2002, Cloix et al.., 2010b). These time periods of MSO-dependent episodes of epilepsy vary from one animal to another, and according to the inbred mouse strains used. Highest augmentation of glycogen leading to high levels of glycogen accumulation takes place as soon as the MSO-induced pre-convulsive period starts which are specifically confined to astrocytes of the cerebral cortex and cerebellum (Hevor et al.., 1985, Hevor et al.., 1986, Hevor, 1994, Bernard-Hélary et al.., 2000, Bernard-Hélary et al.., 2002). In addition, the action of MSO on recurrent alterations in metabolism is also observed *in vitro* in cultured rat and mouse astrocytes. These data show that such metabolic effects occur independent from seizures and are pharmacologically induced well before seizure onset (Hevor and Delorme, 1991, Hevor, 1994, Bernard-Hélary et al.., 2000, Bernard-Hélary et al.., 2002). Alteration of glycogen content is also observed in other models of induced seizures, such as those depending upon homocysteic acid (Folbergrova et al.., 2008), where the accumulation of glycogen was essentially due to astrocytic metabolism. That is also the case in progressive myoclonus epilepsy where inclusion bodies, resembling abnormal glycogen, accumulate during the disease (Wang et al.., 2007). With one exception (Vilchez et al.., 2007) who described that murine neurons posses also the enzymatic machinery to synthesize glycogen, astrocytes are considered to be the only cell types capable for glycogen synthesis (Magistretti and Pellerin, 1996, Pellegri et al.., 1996, Magistretti et al.., 1999, Magistretti, 2006, Cloix et al.., 2008, Cloix and Hevor, 2009).

Some actions of MSO on glutamate receptors, and inhibition of glutamine synthetase resulting in putative accumulation of glutamate, have been reported as being responsible for MSO-dependent seizures (Peters and Tower, 1959, Griffith and Meister, 1978, Eid et al.., 2004, Eid et al.., 2008) whereas others describe a decrease in the brain glutamate content (Fonnum, 1984, Engelsen and Fonnum, 1985, Fonnum and Paulsen, 1990), or no variation (Somers and Beckstead, 1990), after administration of a convulsive dose of MSO in rats, rather than the expected increase. Furthermore, chronic inhibition of brain glutamine synthetase by MSO did not induce seizures in mice (Engelsen and Fonnum, 1985, Blin et al.., 2002), although a unilateral deficiency of hippocampal glutamine synthetase, induced by MSO, causes seizures in a new "pharmacological kindling" model of epileptic rats (Eid et al.., 2008). The precise mechanism responsible for MSO-dependent seizures is therefore not well defined, and further investigations are needed.

The purpose of the present study is to compare two inbred strains of mice, C57BL/6J and CBA/J, and two inbred lines of mice, iMSO-Fast and iMSO-Slow, selected according to their latency toward MSO-dependent seizures (Cloix et al.., 2010b). Indeed, on account of the high resemblance of MSO models to human epilepsy, and because of the recent metabolic data on the associated conditions, we are interested in addressing the basic mechanisms that generated epileptiform activities in our MSO model which may contribute to a better understanding of epileptogenesis. Our approach was based upon selecting mice that responded differently to the administration of MSO, and upon comparing the metabolism and the behavior of selected inbred lines in order to find abnormalities, which can trigger seizures. In preliminary investigations, based upon the first few stages of inbreeding, we observed that MSO basically separated mice into two types, one that developed seizures minutes after MSO injection (MSO-Fast) and secondly those that developed seizures much later after administration

(MSO-Slow) (Cloix and Hevor, 2009, Cloix et al.., 2010a). The comparison of these various strains of mice revealed different sensitivities toward MSO in terms of latency, brain glycogen and neurotransmitter contents. In addition for the MSO-Fast and MSO-Slow inbred lines we investigated effects of various other convulsants and anticonvulsants, as well as the MSO-dependent changes in brain neurotransmitter and glycogen contents as well as altered behavioral characteristics such as anxiety and the capacity for spatial learning. These latter approaches addressed specifically the involvement of glutamatergic and GABAergic pathways in relation to glutamine synthetase activity, neurotransmitter and glycogen content and their putative interrelationships with behavior.

II. Materials and Methods

II.1. Animals

Ten week-old male CBA/J and C57BL/6J mice were obtained from the Transgenèse et Archivage d'Animaux Modèles (TAAM, CNRS UPS44, Orléans, France). MSO-Slow and MSO-Fast mice were selected according to a previously described procedure using latencies toward a 75 mg/kg dose of MSO (Cloix et al.., 2010b, Boissonnet et al.., 2012). Thereafter, 4 additional MSO-challenges were performed in order to increase the selection pressure: MSO-Slow mice were challenged with 100 mg/kg MSO, while MSO-Fast ones were challenged using 50 mg/kg MSO. This was followed by 10-12 inbreeding generations of brother-sister crosses to generate mice (iMSO-Fast, iMSO-Slow) that were 10 week-old at the time of the experiments. Doses of MSO to induce generalized seizures were chosen according to previously described methods, and data describing the different responses toward various MSO-doses (Cloix et al.., 2010b). All protocols were approved by the local ethical committee with the agreement number CREEA CL2007-023, and were in accordance with the European Community Council Directive of 24 November 1986 (86/609/ECC).

Mice were sacrificed by decapitation, and the heads were immediately frozen in liquid nitrogen and thereafter conserved at -80 °C. Dorsal cortices were dissected from the brain at -20 °C and pulverized with a liquid nitrogen frozen homogenizer, and then kept at -80 °C until used. Brain cortices were homogenized in 5 volumes of the appropriate solution (0.2 M $HClO_4$ for glycogen and neurotransmitter measurements; 100 mM imidazol buffer for glutamine synthetase activity). Protein content was determined using the Lowry procedure (Lowry et al.., 1951) with bovine serum albumin as a standard.

II.2. Chemicals and Seizure Latency

Methionine sulfoximine (MSO), pentylenetetrazole (PTZ, an GABA antagonist), kainic acid (KA, an glutamatergic agonist), valproic acid (VPA, an GABA agonist), pilocarpine (PC, an acetylcholine agonist) and MK-801 (an NMDA antagonist) were obtained from Sigma-Aldrich (Saint Quentin Fallavier, France). The chemicals were dissolved in 0.9% NaCl and the pH was adjusted to 7.4 with 1 N NaOH for KA and VPA only. They were administered intraperitoneally (i.p.), and the latencies to seizures were determined as previously described

(Cloix et al.., 2010b). Maximal times of observation were as follows: MSO, 600 min; KA and PC, 120 min; PTZ 180 sec. Mice that did not seize during the observation period were given a score of 600 for MSO, 120 for KA, and 180 for PTZ, and measured latencies included the values corresponding to the time when mice were not convulsing. Doses of various convulsants and anticonvulsants were as follows: MSO, various doses as indicated hereafter and ranging from 0 to 200 mg/kg for MSO-Fast and 0 to 400 mg/kg for MSO-Slow; PTZ, 75 mg/kg; PC, 300 mg/kg; KA, 25 mg/kg; VPA, 250 mg/kg; MK-801, 1 mg/kg. Doses of MSO given to each strain of mice were determined as previously described, and with reference to the sensitivity of each strain for MSO-dependent seizures (Bernard-Hélary et al.., 2000, Bernard-Hélary et al.., 2002, Cloix et al.., 2010b). Such doses corresponded to sub-convulsant doses of MSO. The ED_{50} for MSO were calculated using GraphPad Prism software with a non-linear fitting of data, normalized response and variable slope. In this analysis, only the generalized convulsions were considered. Ten CBA/J and 15 C57BL/6J mice (age 10 weeks) were used. For the experiments with MSO 20 ten-week old mice (10 females and 10 males for each strain) were used.

II.3. Glycogen Measurements

Tissue homogenates were centrifuged at 20,000 × g for 1 h at 4°C, and 100 μl of the supernatants and were buffered at pH 4.8 using 0.2 M sodium acetate buffer. Glycogen in tissue homogenates was cleaved to glucose by α-amyloglucosidase; and the glucose released was measured using a standard glucose oxidase/peroxidase method to obtain total glucose (i.e. free glucose plus glucose from glycogen hydrolysis). Glucose levels were also determined without α-amyloglucosidase hydrolysis to obtain free glucose; glycogen content was calculated by subtraction of total and free glucose levels, as described (Verge et al.., 1996, Cloix et al.., 2010a).

II.4. Measurement of Monoamine Concentrations

Tissue homogenates were centrifuged at 20,000 × g for 1 h at 4°C, and 100 μl of the supernatants and were injected, at least twice, directly into a high-performance liquid chromatograph (HPLC) coupled to an electrochemical detector. Monoamine separations were performed using a Beckman Coulter pump (Gold system 118, Fullerton, California, USA) connected to a Zorbax octadecylsilane column of 4.6 × 250 mm size (Agilent, Palo Alto, California, USA). The electrochemical detector was an amperometric detector (Intro, Antec Leyden, Zoeterwoude, The Netherlands) with a glassy carbon electrode. The mobile phase was run using a Beckman Coulter Pump Gold 118, and was constituted of 100 mM sodium acetate, 20 mM citric acid, 0.4 mM sodium octylsulfate, 0.15 mM disodium EDTA, 8% methanol, pH 5.0. The HPLC was developed using a flow rate gradient summarized in Table 1. The various monoamines were determined, and the retention times were as follows. Tyr (tyrosine), RT=7.28min; DOPAC (3,4-dihydroxyphenylacetic acid), RT=14.12min; NE (norepinephrine), RT=17.68min; MHPG (3-methoxy-4-hydroxyphenylglycol), RT=19.17min; 5-HTP (5-hydroxytryptophan), RT= 23.52min; E (epinephrine), RT=28.63min; 5-HIAA (5-hydroxyindoleacetic acid), RT=29.20min; HVA (homovanillic acid), RT=32.23min; NME (normetepinephrine), RT=32.65min; Trp (tryptophan), RT=36.50min; DA (dopamine), RT=40.97min; ME (metepinephrine), RT=43.23min; and 5-HT (5-hydroxytrypamine or serotonin), RT=57.97min. Concentrations of the different authentic neurotransmitters were injected at least 3 times, and a standard curve was determined for each neurotransmitter, corresponding to the response of the detector as a function of injected contents. Such standard curves were used to determine the amounts of neurotransmitters in the supernatant from brain cortices, which were expressed as pmol/mg protein. For some neurotransmitters, in order to be sure that the obtained peak corresponded to the standard neurotransmitter, the brain samples were over-loaded with known amount of standard neurotransmitter and injected to the system.

Table 1. Chart of gradient flow rates used in the determination of monoamines using HPLC.

Flow (ml/min)	Duration (min)	Starts at (min)	Ends at (min)
0.5	4.0	0.0	4.0
0.1	0.3	4.0	4.3
0.1	19.7	4.3	24
0.4	0.5	24.0	24.5
0.6	9.5	24.5	34.0
0.6	1.0	34.0	25.0
0.4	0.5	35.0	35.5
0.4	10.5	35.5	46.0
2.5	2.0	46.0	48.0
2.5	9.0	48.0	57.0
0.5	2.0	57.0	59.0
0.5	1.0	59.0	60.0

II.5. Glutamine Synthetase Activity

Ten animals (5 females and 5 males) were used to determine kinetic properties of glutamine synthetase (GS) from both inbred strains of MSO-Fast and MSO-Slow mice. Their cortices were pulverized and the powder was processed, as previously described (Blin et al.., 2002), except for homogenization that was performed in 100 mM imidazol buffer. Protein content was determined using the Lowry procedure (Lowry et al.., 1951). Maximal enzyme activity and substrate affinity were determined by increasing amount of substrate, i.e. glutamate from 0.2 to 50 mM; and the inhibition constant was calculated for MSO varying from 1.5 μM to 25 mM, at a glutamate concentration of 50 mM.

II.6. Behavior and EEG

We tested 40 animals, 20 females and 20 males. The radial-maze test, to determine spatial memory learning capacities, was performed using a previously described method (Blin et al.., 2002). Open-field analysis, to determine anxiety, was performed according to a slight modification of previously described protocols (Belzung and Griebel, 2001, Crusio, 2001,

Einat et al.., 2003). In brief, the open-field design corresponded to a dark plastic homemade open-field with 9 square-cases of 15 cm each, separated by a 2 cm wide blue line. A dark plastic wall of 30 cm height surrounded the 9 cases. Mice were positioned in the centre square and observed for 5 minutes. Before each mouse entered the open-field it was wiped with a paper cloth soaked with 70% ethanol and with water thereafter. The following values were recorded by the same experimenter: time before the first movement, corresponding to the time needed to pass to the next square; number of visited squares, which corresponded to the line crossed by at least the animal's shoulders; the number of total rearing; and the number of defecation pellets. These values were considered as an index of the animal's anxiety (Belzung and Griebel, 2001, Crusio, 2001, Einat et al.., 2003). Electroencephalogram (EEG) recording was performed as follows. Under isoflurane anesthesia (1.5%), mice were implanted with three monopolar surface electrodes placed in the cranial bone. Two electrodes were set bilaterally over the parietal cortex and a ground electrode was placed over the frontal cortex. Electrodes were made of a tungsten wire (diameter 250 μm) soldered to a male connector (Wire-pro, Farnell, Villefranche sur Saône, France). They were inserted in the skull so that only the tip (0.5 mm) protruded onto tissue. The electrodes were glued to the skull with cyanoacrylate and dental cement. EEG was recorded on freely moving mice, after a recovery period of 7-10 days. EEG was monitored for a several hours lasting period (8 h) using a Powerlab 26T model ML856 ADInstruments with a LabChart v7 sofware (Oxford, UK). The behavior of the mice was observed during the same period of time.

II.7. Cell Culture of Astrocytes and Cell Surface Glycoprofilling

II.7.1. Astrocyte Culture

Astrocytes were cultured according to a previously described method (Bernard-Hélary et al.., 2002). The MSO dose of 10 mM and the duration of incubation had been previously reported (Bernard-Hélary et al.., 2002). For the detection of the astrocytes that were recognized by lectins, these astrocytes were harvested by treatment for 5 min with PBS containing trypsin (250 μg/ml) and EDTA (200 μg/ml). After harvesting, cells were washed in PBS and collected by centrifugation. The resulting pellet was resuspended at 2×10^6 cells/ml and stained with 20 μM diacetylated N-succinimidyl-carboxyfluoresceine (CFSE) for 15 min at 37 °C. After washing with PBS, cells were resuspended at 2×10^6 cells/ml in PBS supplemented with 1 mM $CaCl_2$ and 0.5 mM $MgCl_2$.

II.7.2. Preparation of Lectin Arrays

Lectins are presented as function of their inhibitive saccharide properties. *N*-acetyl-D-galactosamine: BPA (*Bauhinia purpurea* agglutinin) and WFA (*Wisteria floribunda* agglutinin); galactose: MPA (*Maclura pomifera* agglutinin), AIA (*Artocarpus integrifolia* agglutinin) and RCA-I (*Ricinus communis* agglutinin-I); *N*-acetyl-D-glucosamine: GSL-II (*Griffonia simplicifolia* lectin-II); sialic acid: MAA (*Maackia amurensis* agglutinin); complex structures: PHA-E (*Phaseolus vulgaris* agglutinin-E). Lectins were directly printed in triplicates on Biomat (Rovereto, Italy) fluorescent black 96 wells plates according to GLYcoDiag technology (Landemarre et al.., 2013, Landemarre and Duverger, 2013).

II.7.3. Application to Lectin Array and Data Processing

One hundred µl (about 2 x 10^5 cells) were incubated for 2h at 37 °C under gentle agitation (25 rpm). After washing, fluorescence intensity was measured using Alpha Fusion universal microplate analyzer (Packard).

II.8. Statistical Analyses

Data are expressed as mean ± standard error of the mean (SEM) for the numbers of animals indicated. The results of the latency toward MSO, various convulsants and anticonvulsants, and glutamine synthetase activity were analyzed using Student *t*-test. The results for glycogen and neurotransmitter variations and for radial-maze were evaluated by repeated measures of ANalysis Of VAriance (ANOVA), with STRAIN (MSO-Slow *vs* MSO-Fast) and SEX as between-subjects factors, and DAYS as within-subject factor. Strain and sex comparisons for individual days and convulsion were made by means of two-way ANOVAs. As some animals submitted to radial-maze did not always take the food reward the first time they entered an arm on days 1 and 2, only the data from days 3-5 were used for the statistical evaluation. The results of open-field trials were evaluated by repeated measures of ANOVA, with STRAIN (MSO-Slow *vs* MSO-Fast) and SEX as between-subject factors. Data was considered significant when the *p* value was less than 0.05%.

III. Results

III.1. MSO Induced Various Latencies in the Four Strains of Mice

When MSO was i.p. administered to various mouse strains as a unique dose latencies toward MSO-dependent generalized seizures varied, as exemplified in Figure 1. Although the convulsant doses were not the same, a significant difference in latencies was observed between CBA/J and C57BL/6J mice, and between MSO-Fast and MSO-Slow mice. The latency period towards MSO-dependent seizures was significantly increased in C57BL/6J than that in CBA/J, and in MSO-Slow when compared to MSO-Fast. Moreover, latencies obtained after MSO administration were significantly different from each other, as determined using ANOVA analysis (Figure 1A). The percentage of convulsing mice was differed among the four stains of mice. The order of the decreasing latencies of convulsing mice after MSO administration was MSO-Slow, C57BL/6J, CBA/J, and MSO-Fast, the latter being the fastest responder toward MSO (Figure 1B). These data were in agreement with the ED_{50} of MSO that induced seizures in 50% of mice (Figure 1C). The most resistant strain was MSO-Slow as the ED_{50} of MSO was not computable, since only 25-30% of mice seized when given large MSO-doses such as 300-500 mg/kg. The ED_{50} of MSO was significantly different between CBA/J and C57BL/6J, and between MSO-Fast and MSO-Slow. An increasing ED_{50} of MSO was observed in the following order: MSO-Fast ≈ CBA/J < C57BL/6J < MSO-Slow. In addition, ED_{50} values obtained after MSO administration were significantly different from each other, as determined by ANOVA analysis (Figure 1C). In addition, it should be mentioned that no seizure was observed during the test duration of 10 hours after administration of 75 mg/kg MSO to MSO-Slow inbred mice. These data strongly suggested that MSO-Slow were really resistant to MSO-dependent seizures whereas MSO-Fast ones were highly sensitive.

C57BL/6J and CBA/J showed an intermediate responsiveness toward MSO indicating that C57BL/6J the more resistant one of the two. Moreover, the data revealed that the selection of the MSO-Fast and MSO-Slow strains (Cloix et al.., 2010a, Cloix et al.., 2010b) was efficient enough to produce two strains each with a different, extreme responsiveness toward MSO-induced seizures. These data showed that the selection of inbred iMSO-Slow and iMSO-Fast provided experimental animals to address various modes of action of MSO in a more pronounced manner compared to other animal models as intended.

III.2. MSO Induced a Large Increase in Brain Glycogen Content in the Four Strains of Mice

The different time courses observed for the increasing glycogen content in cerebral cortices of mice after an injection of a single dose of MSO are presented in Figure 1D. In all strains, glycogen increased significantly as a function of time. However, the observed increase differed in a strain-specific manner. Glycogen increased to significantly higher levels in C57BL/6J than in CBA/J and it was significantly higher in MSO-Slow than in MSO-Fast mice. Moreover, glycogen variation showed no difference between CBA/J and MSO-Fast mice, the latter two containing significantly lower glycogen levels compared to both MSO-Slow and C57BL/6J.

The presented data suggested that, in terms of cerebral glycogen accumulation, the two strains of mice, C57BL/6J and CBA/J, are intermediate to the two selected strains, MSO-Fast and MSO-Slow. Comparing the obtained data for the glycogen content with the one observed for seizure responsiveness (Figure 1A and 1B), it became obvious that increasing amounts of glycogen correlated with longer latency periods towards MSO-induced seizures, as previously hypothesized (Bernard-Hélary et al.., 2000, Belzung and Griebel, 2001, Crusio, 2001, Bernard-Hélary et al.., 2002, Einat et al.., 2003).

Therefore, as CBA/J and C57BL/6J are in between MSO-Fast and MSO-Slow for the above reported effects of MSO, it might be more than relevant to obtain inbred MSO-Fast and MSO-Slow mice. Thus, we decided to further enhance the diverting responsiveness toward MSO-induced effects of the selected MSO-Fast and MSO-Slow strains by an additional sequence of inbreeding (Fig. 2). The comparison of the responses of these two new "hyperinbred" lines designated as iMSO-Fast and iMSO-Slow with that of their respective parent strains was expected to provide a better understanding of the causes why animals respond differently to MSO- or pharmacologically-induced seizures.

III.3. Selection of MSO-Fast and MSO-Slow Mice

The process of selection of MSO-Fast and MSO-Slow mice was schematized on Figure 2 (Cloix et al.., 2010a, Cloix et al.., 2010b, Boissonnet et al.., 2012). The G0 was generated employing a genetic mixture of 8 different inbred strains of mice. During the 6 first steps in the selection process each mouse was challenged with 75 mg/kg MSO. Mice with the shortest latencies were inter-crossed to lead to a MSO sensitive line, or MSO-Fast mice as shown in

Fig. 2. Conversely, mice with the longest latency period were inter-crossed to give rise to the MSO resistant line, or MSO-Slow mice (Fig. 2). After these first 6 challenges, the selection pressure was increased, and MSO-Fast mice and MSO-Slow mice were challenged using 50 and 100 mg/kg MSO, respectively. After these two steps, during which the crosses between sister and brother were avoided and the response toward MSO was considered as obtained, the process of inbreeding was started for at least 12 crosses between sister and brother only. Afterward, the obtained iMSO-Fast and iMSO-Slow mice were considered as inbred.

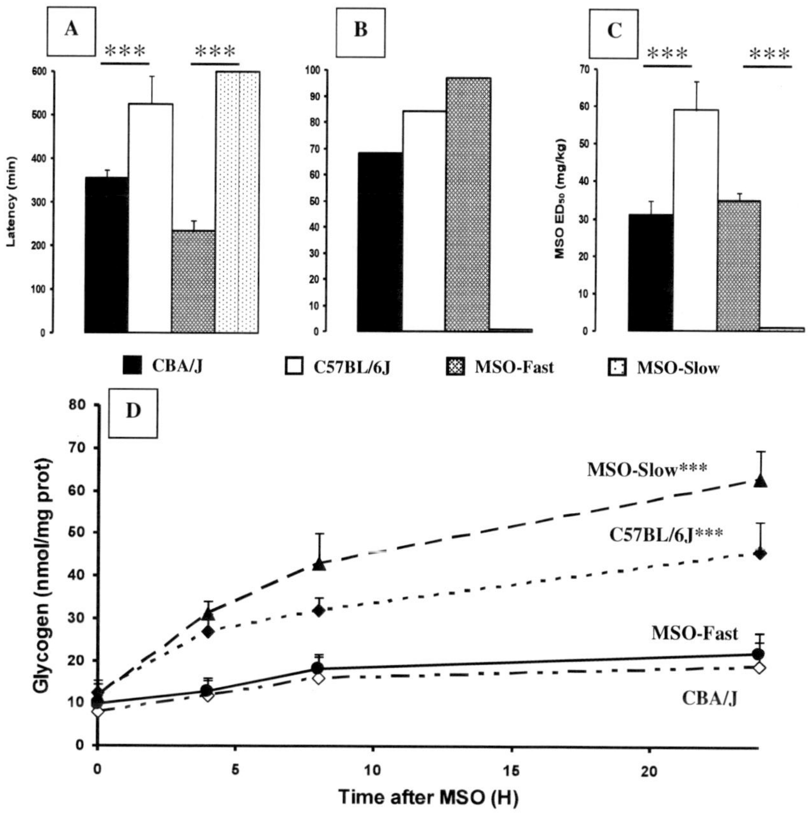

Figure 1. Characteristics of MSO-dependent seizures within two inbred mouse strains, and MSO-Fast and MSO-Slow after G6. MSO was administered using intraperitoneal injection as a unique dose to different inbred strains of mice of both sexes: CBA/J, 40 mg/kg, n=10; C57BL/6J, 100 mg/kg, n=15; MSO-Fast, 75 mg/kg, n=20; MSO-Slow, 250 mg/kg, n=20. A: Latency toward MSO determined as time-duration between MSO injection and the first generalized seizure. When mice did not seize during the 10 hour period of observation a score of 600 was assigned. B: Percentage of mice that seized. C: ED_{50} of MSO was determined as previously described (Bernard-Hélary et al.., 2002). NC: no computable value using GraphPrims software. Values are expressed as mean ± SEM, and significance between two inbred mice, C57BL/6J *vs* CBA/J, MSO-Fast *vs* MSO-Slow, was assumed using Student *t*-test as p<0.001: ***. D: Variation of brain glycogen content in the four strains of mice. Mice were given an unique dose of MSO according to the strain: CBA/J, 40 mg/kg, n=10; MSO-Fast, 75 mg/kg, n=20; C57BL/6J, 100 mg/kg, n=15; MSO-Slow, 75 mg/kg, n=20. Mice were sacrificed by decapitation at various times as indicated and glycogen was determined as previously described in materials and methods (Cloix et al.., 2010a). Values are expressed as mean ± SEM, and significance between glycogen variation as a function of time in inbred mice was tested using repeated ANOVA test as *p*<0.001: ***.

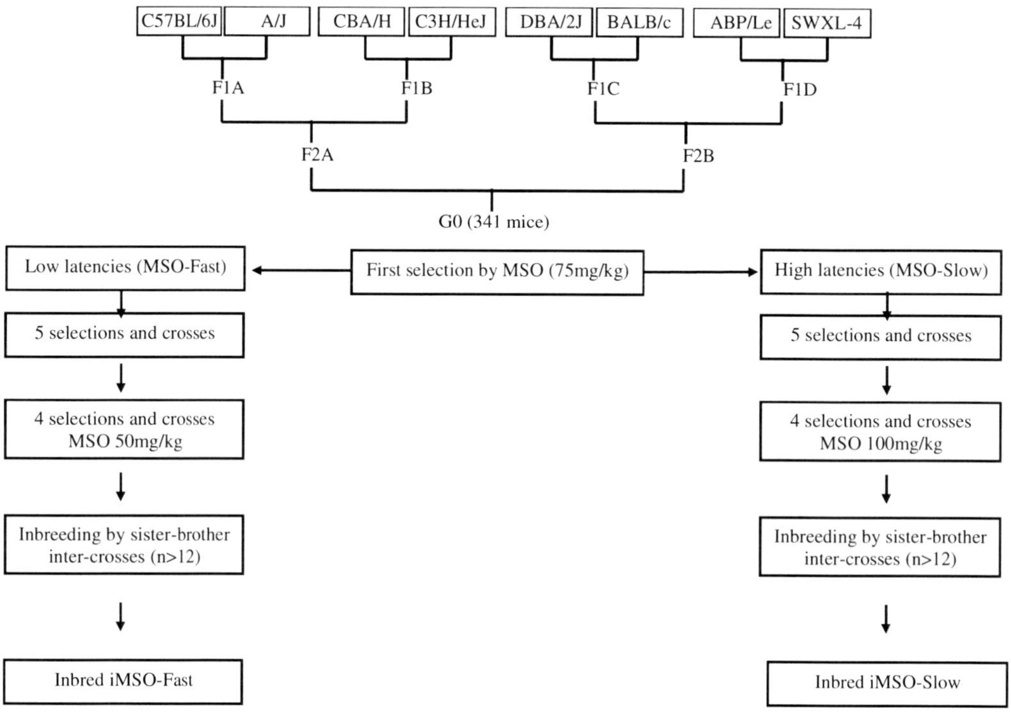

Figure 2. Selection of MSO-Fast and MSO-Slow inbred mice based upon MSO-dependent seizures latencies.

These iMSO-Fast and iMSO-Slow mice after inbreeding were characterized and in part compared to their respective ancestors (MSO-Fast; MSO-Slow, G6 = sixth generation) in terms of certain parameters specified below.

III.4. Response Toward MSO of iMSO-Fast and iMSO-Slow Mice

III.4.1. Latency and Seizures

The MSO-dependent seizure responses were analyzed in both strains, iMSO-Fast and iMSO-Slow during and after the selection process, and are exemplified in Figure 3. In terms of latencies, iMSO-Slow mice were significantly lower responders or no-responders toward MSO than other lines of inbred mice (Figure 3A). These data were similar to those obtained when only mice that seized were considered (Figure 3B). The reported data confirmed the usefulness of the two inbred strains, iMSO-Fast and iMSO-Slow. These inbred mice offered the potential for studying the mechanisms involved in MSO-dependent seizure induction.

These data were corroborated by the EEG measured in inbred MSO-Fast and MSO-Slow mice (Figure 4) receiving 75 mg/kg MSO to MSO-Fast and 350 mg/kg MSO to MSO-Slow. The MSO-Fast indeed responded few alterations in EEG 2 hours after MSO administration which were higher than those observed at the same time point in MSO-Slow. Moreover, 4½ hours after MSO injection EEG was deeply altered in MSO-Fast mice whereas a more moderate modification could be observed in inbred MSO-Slow mice.

Altogether, these data demonstrated that the selection process indeed produced two different inbred lines of mice, which differed in terms of their seizure responses toward MSO

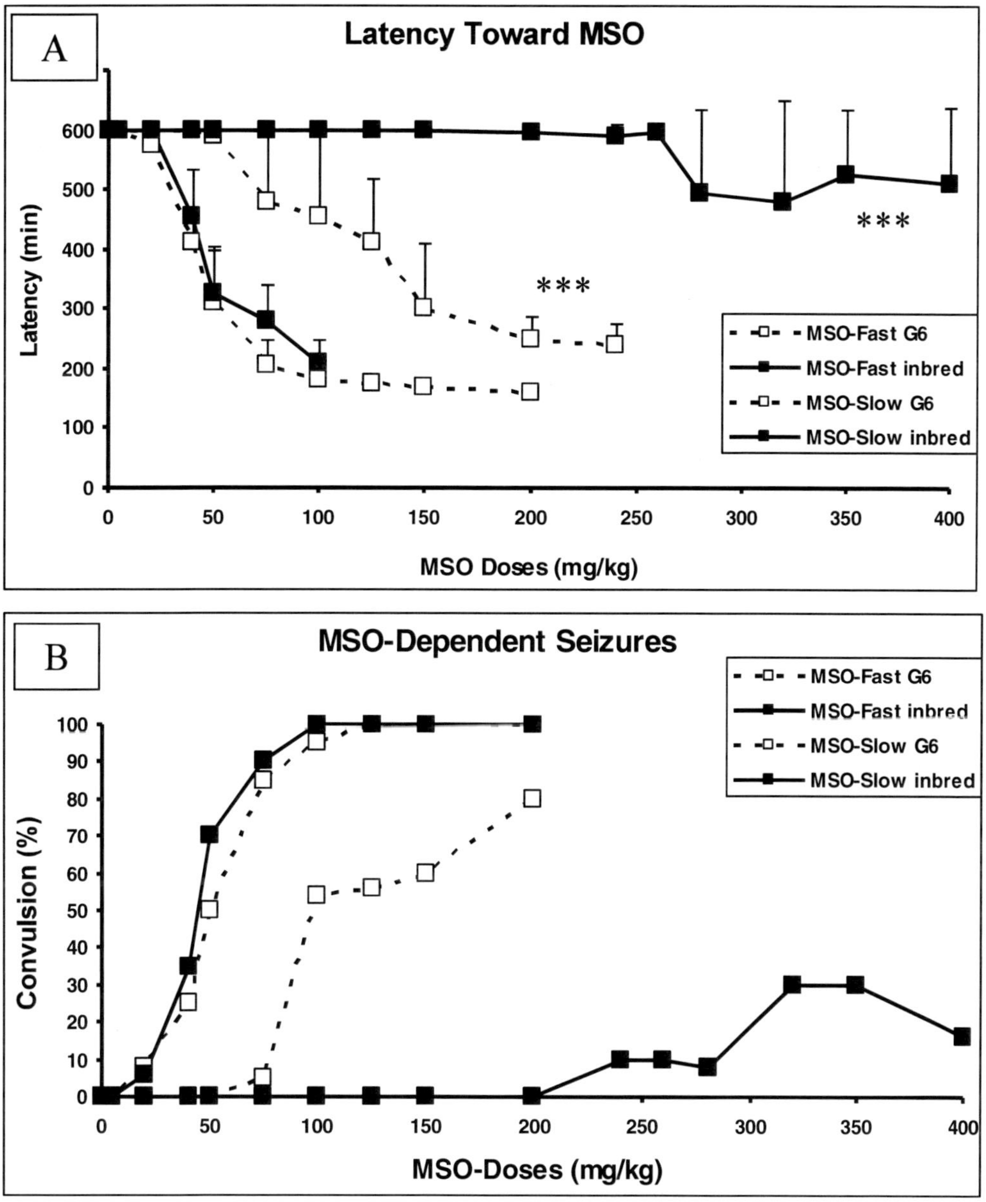

Figure 3. Characteristics of MSO-dependent seizures of MSO-Fast and MSO-Slow of both sexes (20 mice, 10♂, 10♀). Used mice corresponded to either G6 mice or inbred mice, i.e., iMSO-fast and iMSO-Slow mice. A: Latencies toward various doses of MSO administered i.p. Latencies were determined as time period between MSO injection and the first observed generalized seizure. Mice which did not seize were arbitrarily given a score of 600. ***: p<0.001 using ANOVA analysis. B: Seizures were expressed as percentage of seizing mice.

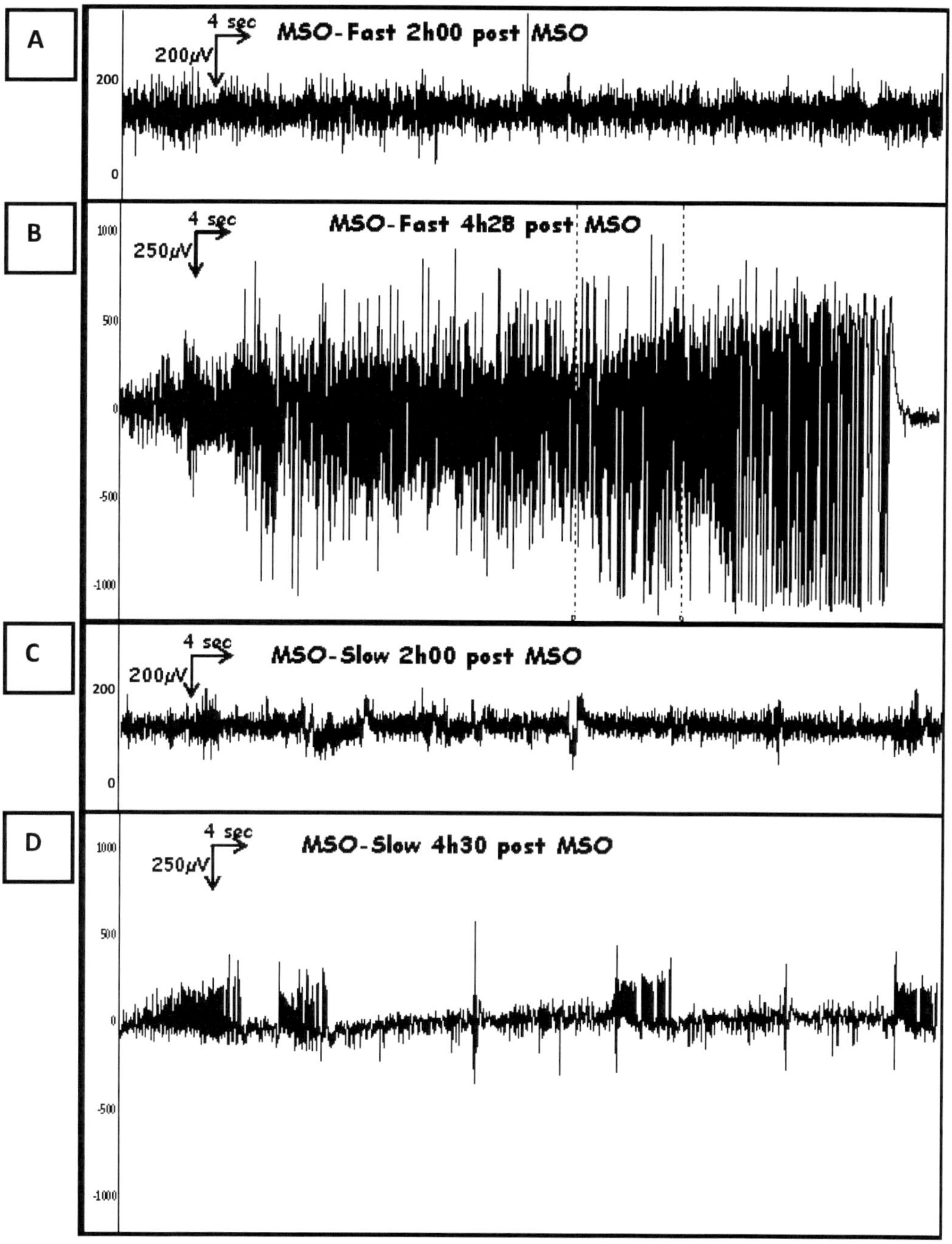

Figure 4. Typical electroencephalogram for MSO-Fast (A, B) and MSO-Slow mice (C, D) of both sexes at G6. EEG was continually recorded according to the method described either 2 hours after 75 mg/kg MSO (A, C) or 4½ hours after 350 mg/kg MSO (B, D). MSO-Fast and MSO-Slow mice were challenged with a dose of MSO derived from data presented Figure 3.

III.4.2. Responses of iMSO-Fast and iMSO-Slow to Other Convulsants and Anticonvulsants

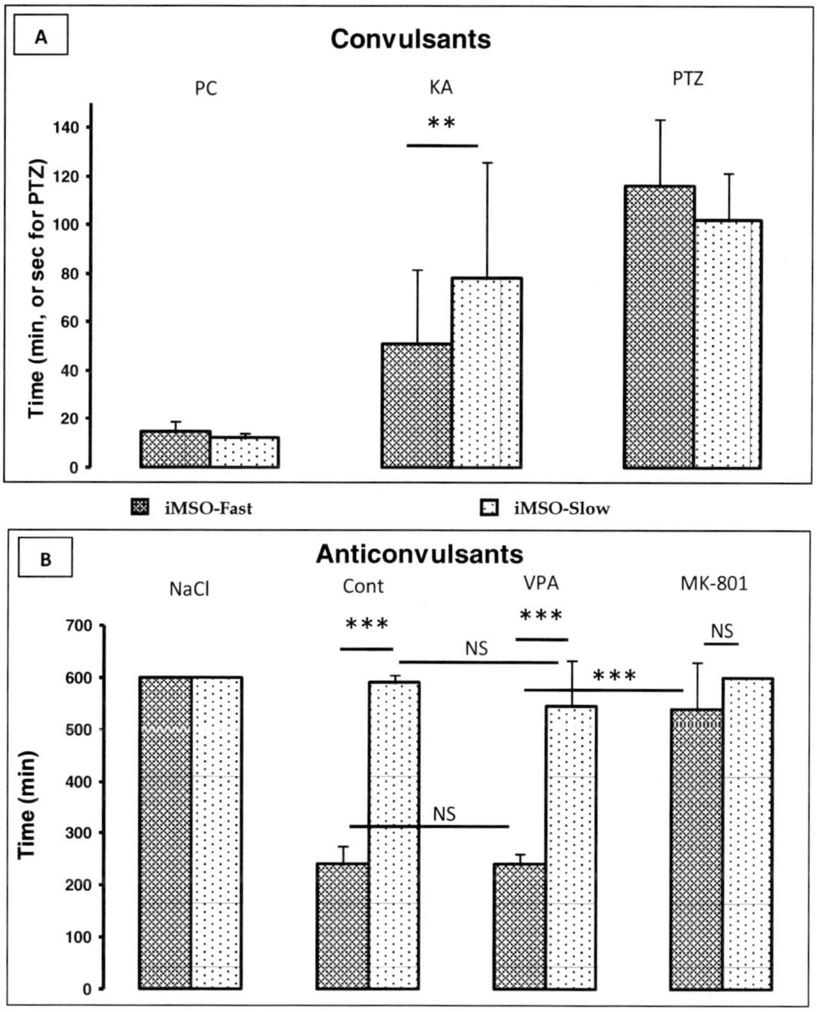

Figure 5. Latency towards various convulsants (A) and effect of two anticonvulsants (B) in MSO-Fast and MSO-Slow mice (n=20, 10 ♂, 10 ♀). MSO was i.p. administered to iMSO-Fast and iMSO-Slow at doses of 75 and 350 mg/kg, respectively. A: Mice were given convulsants pilocarpine (PC, 300mg/kg), kainic acid (KA, 25mg/kg), pentylenetetrazole (PTZ , 75mg/kg). The latency was determined and mice not convulsing were given a score of 60 minutes for pilocarpine, 120 minutes for kainic acid, and 180 seconds for PTZ. B: Thirty minutes before MSO mice were given saline (cont), or MK-801 (1mg/kg) or valproic acid (VPA, 250mg/kg). The latency was determined and mice not convulsing were given a score of 600 minutes for MSO (NaCl, given instead of anti-convulsant). Values are expressed as mean ± SEM, and significance between latencies in MSO-Fast and MSO-Slow inbred mice was determined using Student *t*-test as $p< 0.05$: *; $p<0.001$: ***.

Figure 5 shows for iMSO-Fast and iMSO-Slow mice their responses towards other convulsants as well as their responses to anti-convulsants after a convulsive dose of MSO. The only convulsant, other than MSO, which demonstrated a significant difference between the two inbred MSO-strains was kainic acid. Latencies toward seizures induced by kainic acid were indeed longer in iMSO-Slow mice than in iMSO-Fast. Moreover, only MK-801 was able to significantly antagonize the MSO-dependent seizures in iMSO-Fast mice. From these data, we suggested that MSO might act through the glutamatergic pathway rather than the GABAergic one. These data were consistent with the putative mechanism of action for MSO as previously reported (Meister, 1978, Eid et al.., 2008, Perez et al.., 2012).

III.4.3. Brain Glutamine Synthetase Activity

Since the data indicated a preferential involvement of the glutamatergic pathway during the selection process of MSO-Fast and MSO-Slow mice, the activity of the main enzyme responsible for transforming glutamate to glutamine, characteristics of glutamine synthetase (GS), were evaluated in cerebral cortices of both MSO-Fast and MSO-Slow mice (Figure 6). V*max* (Figure 6A) was therefore determined by increasing the glutamate concentrations, and in this condition the V*max* for GS from iMSO-Fast was significantly higher compared to that of iMSO-Slow. When Km values for glutamate and K*i* for MSO were addressed also (Figure 6B), no differences were observed between iMSO-Fast and iMSO-Slow mice.

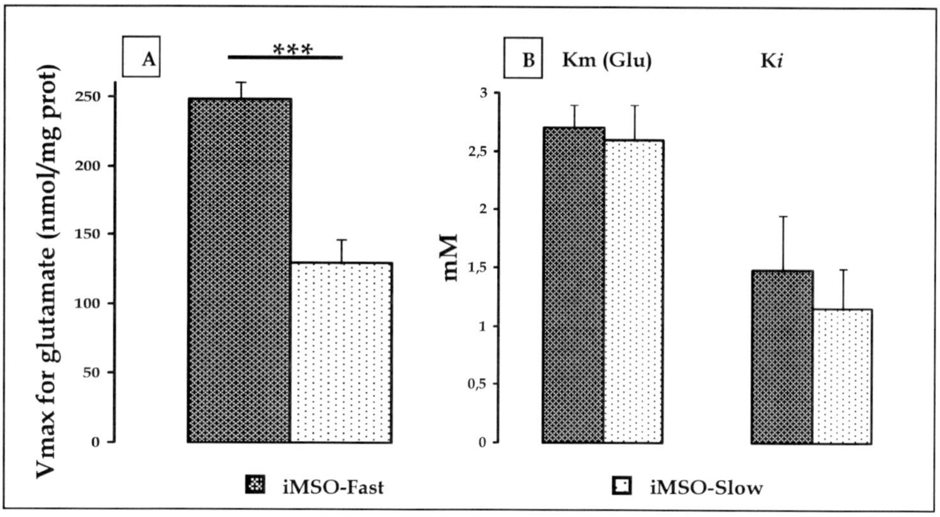

Figure 6. Glutamine synthetase (GS) activity of cerebral cortices of iMSO-Fast and iMSO-Slow mice. GS activity was determined in cortices of both inbred mice as previously described (Blin et al.., 2002). A: V*max* values were determined with increasing concentrations of glutamate ranging from 0.2 to 50 mM. B: Determination of Km value for glutamate using varying glutamate concentration from 0.2 to 50 mM, and K*i* value for MSO using increasing concentration of MSO from 1.5 μM to 25 mM with saturating glutamate concentration of 50 mM. Values are expressed as mean ± SEM and significance between iMSO-Fast (n=10) and iMSO-Slow (n=10) inbred mice of both sexes, was assumed using Student *t*-test as $p<0.001$: ***.

These data suggested either an intrinsically higher GS activity, or a higher level of enzyme expression in the iMSO-Slow lines compared to iMSO-Fast ones. Similarly, since no significant difference for GS-inhibition by MSO was observed between the two strains of mice, it could be concluded that the sensitivity of GS toward MSO might not be different among the strains and was therefore not responsible for the differing responses towards MSO-induced seizure activity. Moreover, since the activity of GS is significantly different between the two strains of mice, the glutamate/glutamine cycle may be involved in the process of MSO-dependent seizures within.

III.4.4. MSO Induced Variation in Neurotransmitter Contents in Brain Cortices

The above results suggested that the observed differences in GS-activity may lead to different concentrations of glutamine derived neurotransmitters such as Glu and GABA as well as others; we therefore analyzed their cerebral contents in both iMSO-Fast and iMSO-Slow mice either after MSO administration or under control conditions (Figures 7 and 8).

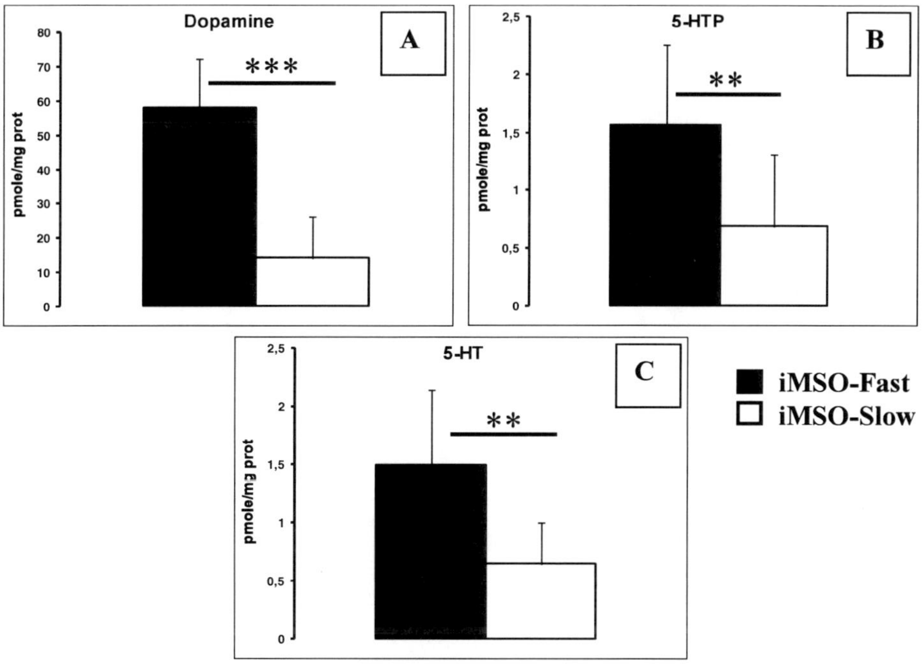

Figure 7. Variation of brain cortex basal content of DA (A), 5-HTP (B) and 5-HT (C) in inbred iMSO-Fast and iMSO-Slow mice. Mice were sacrificed by decapitation and neurotransmitters were determined using a reverse-phase HPLC coupled to an amperometric detection as previously described in the section of materials and methods (Cloix et al.., 2010a). Values are expressed as mean ± SEM, and significance was assumed *t* test $p<0.01$: **; $p<0.001$: ***, for n=8, (4 ♂ and 4 ♀).

Basal levels of cerebral neurotransmitters (Figure 7) demonstrated significantly higher levels in iMSO-Fast than in iMSO-Slow for DA (Figure 7A), 5-HTP (precursor of 5-HT; Figure 7B) and 5-HT (Figure 7C) only. The cerebral levels of other neurotransmitters were not significantly changed. Therefore, in basal conditions, cerebral contents of DA, 5-HTP, and 5-HT itself were significantly increased in iMSO-Fast mice. The high 5-HT content might be relevant in iMSO-Fast mice since 5-HT had been reported as an anti-convulsant (Blizard and Balkoski, 1982) and described as modified in epileptic models (King and LaMotte, 1989, Pranzatelli and Nadi, 1995). We hypothesize therefore that 5-HT was increased in MSO-Fast mice in order to cope with a higher risk of hyperexcitation even under normal conditions.

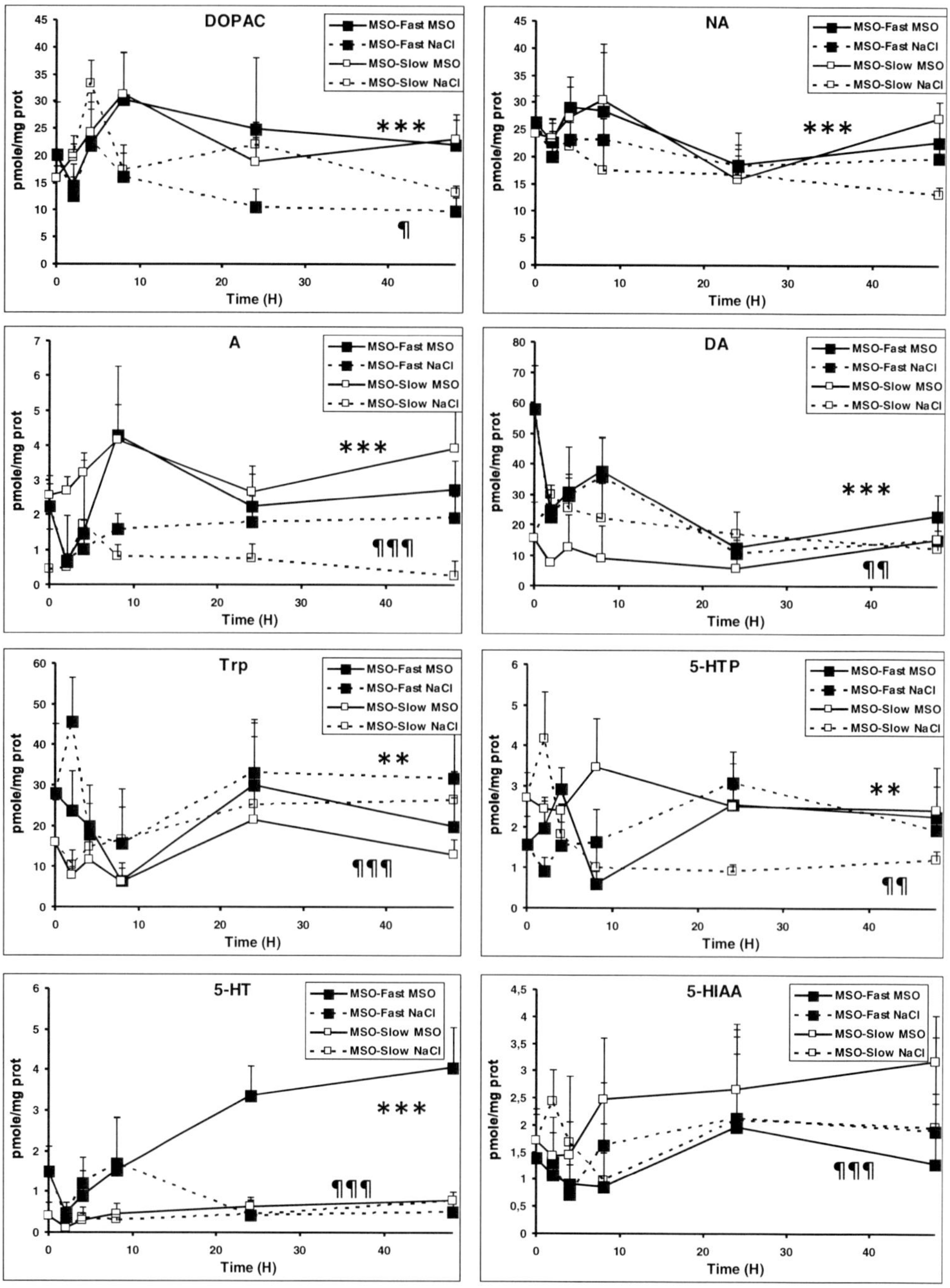

Figure 8. Variation of brain cortex content of monoamines in inbred iMSO-Fast and iMSO-Slow mice after administration of either 75 mg/kg MSO (n=10) or NaCl (n=10). Mice were sacrificed by decapitation at various indicated time and neurotransmitters were determined using an HPLC coupled to an amperometric detection as previously described in the section of materials and methods (Cloix et al.., 2010a). Values are expressed as mean ± SEM, and significance was assumed using repeated ANOVA test: $p<0.01$: **; $p<0.001$: *** when considering effect of treatment, i.e. MSO *vs* NaCl, in both inbred lines; and $p<0.05$: ¶; $p<0.01$: ¶¶; $p<0.001$: ¶¶¶ when considering differences between iMSO-Fast and iMSO-Slow, in both treatments.

The significant effects on neurotransmitter changes after challenging both inbred MSO-strains with MSO compared to saline are shown in Figure 8. Time variation of DOPAC (3,4-dihydroxyphenylacetic acid), NA (norepinephrine), A (epinephrine), DA (dopamine), Trp (trytophan), 5-HTP (5-hydroxytryptophan) and 5-HT (5-hydroxytrypamine or serotonin) were significantly increased by MSO treatment when compared to control mice in both inbred strains. In terms of difference in the variation of monoamines between iMSO-Fast and iMSO-Slow mice, a significant difference between the two strains was observed for DOPAC, A, DA, Trp, 5-HTP, 5HT and 5-HIAA (5-hydroxyindoleacetic acid). The MSO-induced changes of these monoamines were higher in iMSO-Fast mice than in iMSO-Slow ones, except for Trp and DA which underwent similar alterations in both strains.

All these results indicated that the serotinergic pathway was more activated in iMSO-Fast mice when compared to iMSO-Slow, whereby MSO induced an increase in both strains. Taken together it suggested that the basal synthesis and utilization of 5-HT is enhanced in iMSO-Fast mice compared to iMSO-Slow. The latter may imply a role for a 5-HT related mechanisms which became activated in order to cope with MSO-dependent seizures.

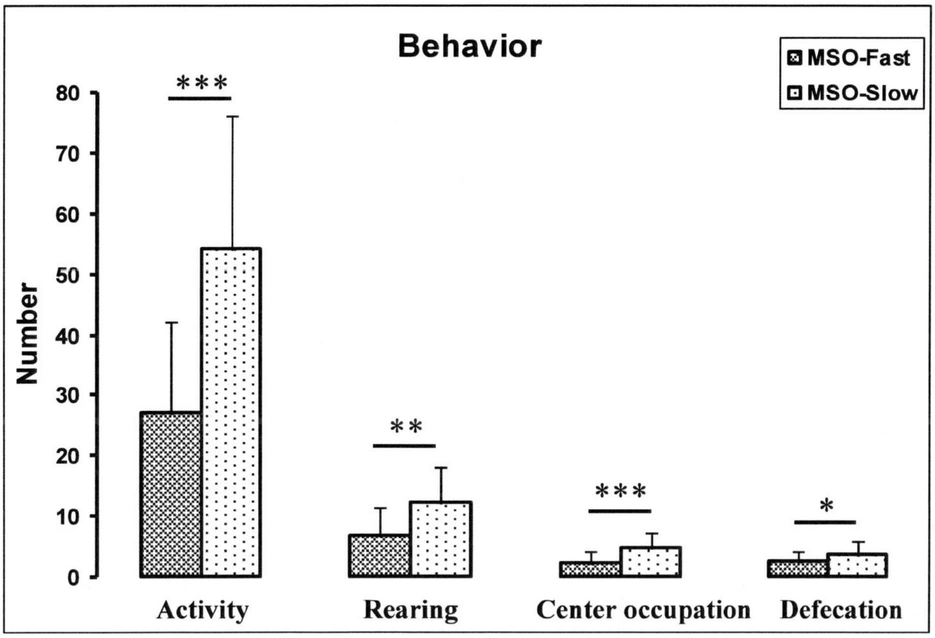

Figure 9. Behavior of iMSO-Fast and iMSO-Slow mice. Mice were put in the central square of the open-field and observed for 5 minutes. Activity corresponded to the numbers of crossed lines by the shoulder of mice. Center occupation corresponded to the numbers of return to the central square. Defecation corresponded to the number of feces pellet. Values are expressed as mean ± SEM and significance between iMSO-Fast (n=40, 20 ♂, 20 ♀) and iMSO-Slow (n=40, 20 ♂, 20 ♀) inbred mice, was assumed using Student *t*-test as $p<0.05$: *, $p<0.01$: **, $p<0.001$: ***.

III.4.5. Behavioral Studies

The spatial memory and the anxiety of iMSO-Fast and iMSO-Slow mice were compared. No significant differences between the selected strains could be observed in terms of spatial learning capacities (data not shown), and the data on anxiety analyzed by the open-field test are presented in Figure 9.

In general, iMSO-Slow mice were more active than iMSO-Fast, which had been validated by the significantly lower values, observed for iMSO-Fast than for MSO-Slow, in terms of number of squares visited and rearing, the time spent in the central case before mice entered the neighbor case, and the number of defecation determined in the open-field (Figure 9). These values were considered as an index of anxiety for mice (Belzung and Griebel, 2001, Crusio, 2001, Einat et al.., 2003). It seems quite obvious that iMSO-Fast mice were more anxious than the MSO-Slow. This might not be directly related to acute MSO-dependent effects and could be a consequence of the selection process. In the latter case, MSO challenging in the course of producing these two strains (MSO-Fast and MSO-Slow) the strain which suffered more may have already inherited a phenotypic more pronounced anxiety behavior which has to be addressed in future studies especially to obtain information on whether epigenetic factors are related to this phenomenon.

Our data were in agreement with reports underlining the involvement of the glutamatergic system in learning capacity and in anxiety. Therefore, we can assume that the data indicating higher GS activity in iMSO-Slow than in iMSO-Fast could possibly contribute to a higher level of anxiety in MSO-Fast as compared to MSO-Slow (Mohler, 2006, Hashimoto et al.., 2007, Venault and Chapouthier, 2007b, Kanner, 2009, Duvoisin et al.., 2010a, Duvoisin et al.., 2010b). Furthermore, for MSO-induced seizures in rats, focal changes in neurotransmitters binding have been described for hippocampal formation as well as for the amygdala, the latter of which is a key structure within the cerebral anxiety circuitry (Cremer et al.., 2009, Cremer et al.., 2010, Pape et al.., 2010, Pape and Pare, 2010) which may indicate that even during the selection process of the strains alterations within the amygdala may have led to differences observed in these two inbred strains.

III.4.6. MSO Induced Variation in Glycogen Contents in Cerebral Cortices

Basal glycogen content and variation of glycogen as a function of time are shown in Figure 10. The basal content of glycogen in cerebral cortices (Figure 10A) was significantly elevated in iMSO-Fast mice compared to iMSO-Slow. Consequently, the variation of brain glycogen content after MSO administration is higher in iMSO-Fast mice than in iMSO-Slow (Figure 10B). In parallel, glycogen content of iMSO-Fast and iMSO-Slow mice did not change after saline administration. These data demonstrated that in iMSO-Fast mice the glycogen level as well as the changes in the level of glycogen in response to MSO was higher compared to iMSO-Slow mice. Therefore, in iMSO-Fast higher amounts of cortical glycogen could theoretically be used to support neurons during their more pronounced seizure activity. iMSO-Fast mice could use the elevated glycogen-produced G-6-P to provide neurons with directly utilizable metabolites to support their energy demand. Conversely, iMSO-Slow either could not provide neurons with energy or the use of glycogen-derived G-6-P was more rapidly utilized to postpone seizures. However, it should be recognized that in iMSO-Fast mice the basal cerebral amount of glycogen was increased when compared to their ancestral generation MSO-Fast G6 (Figs. 1 and 10). Therefore our data do not support the previously reported view that increasing amounts of glycogen may be directly involved in a prolongation of the latency period. This is because latency toward MSO-induced seizures was significantly shorter in iMSO-Fast despite their high basal glycogen content and their more pronounced glycogen enhancement iMSO-Fast mice provide clear evidence that higher glycogen content on its own is not directly involved in the suppression of mechanisms involved in seizure generation.

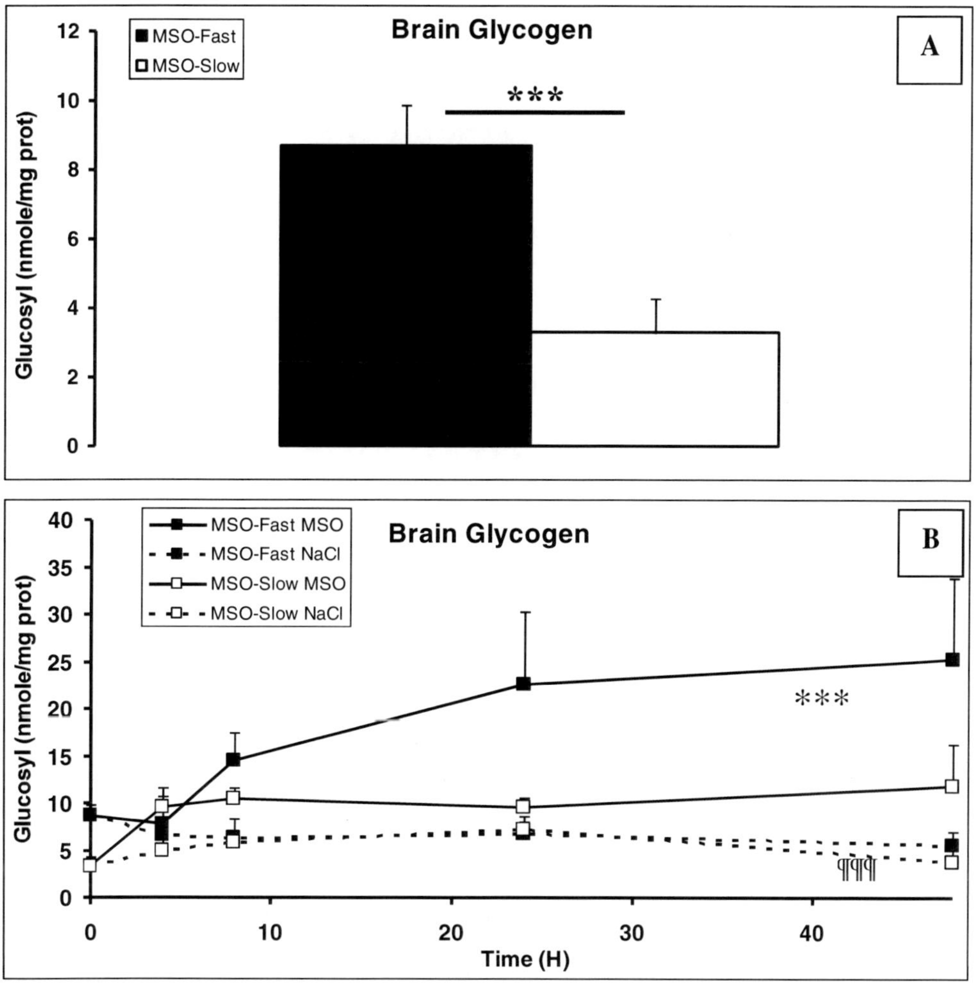

Figure 10. Brain glycogen content determined in basal conditions (A) and after administration of either 75 mg/kg MSO of saline (B) to iMSO-Fast and iMSO-Slow mice. Values are expressed as mean ± SEM (n=10), and significance was assumed using either Student *t* test (A) or repeated ANOVA test (B), when considering effect of treatment, i.e. MSO *vs* NaCl, in both inbred lines: $p<0.001$: ***; and when considering effect of ID, i.e. MSO-Fast *vs* MSO-Slow, in both treatments: $p<0.001$: ¶¶¶, respectively.

III.5. MSO Induced Variation upon Astrocyte Cell Surface in Cerebral Cortices

Astrocytes are able to accumulate glycogen in the brain which could be transformed down to G-6-P and various other glucose metabolites. Xenobiotics such as MSO sometimes modify signaling thereby inducing major changes in cells. As glycosylation is an important way to cell signaling we tried to evaluated the glycosylation of astrocyte membrane proteins in order to look for a possible involvement in altered astrocyte metabolism. For this cell membrane glycans of primary astrocyte cultures were analyzed by glycoprofiling (Figure 11). Data were standardized using the PHA-E signal as reference, and the effect of 10 mM MSO was analyzed 7 hours after its addition to the cultures. The obtained ratios for the interaction between astrocytes treated with 10mM MSO or saline is represented Figure 11. MSO induced higher interactions (ratio >100%) for astrocytes derived from iMSO-Slow mice with WFA, MPA, AIA and MAA lectins whereas a higher ratios of interaction (ratio>100%) from astrocytes of iMSO-Fast mice were observed with WFA, AIA and GSL-II. These data indicated that the above listed lectins recognized glycans which could be involved in MSO-induced changes in the processing of astrocytic cell membrane glycoconjugates. Therefore, we hypothesize that MSO affects extracellular matrix components and related intercellular communication the latter of which could also be contributing to the induction of seizures.

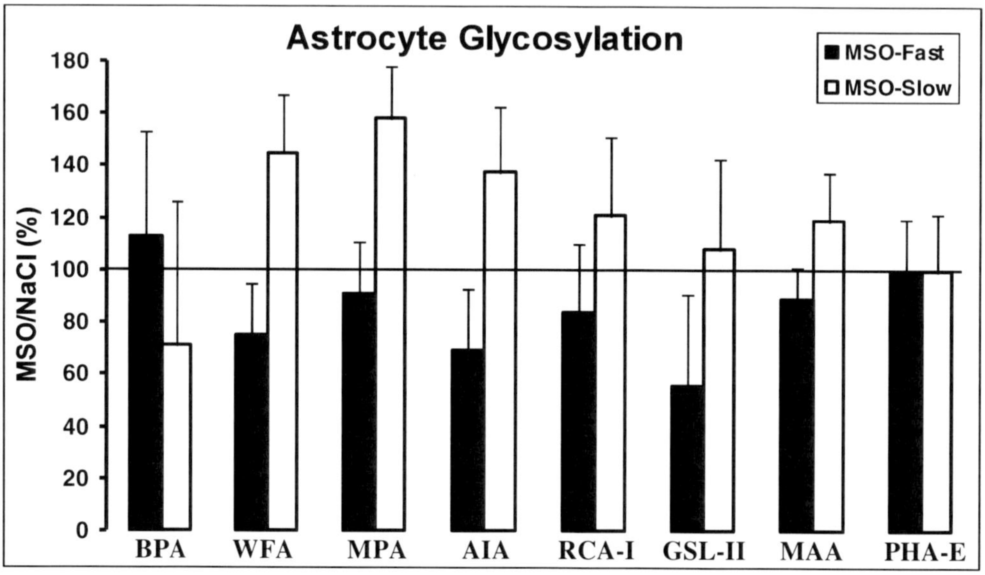

Figure 11. Glycosylation profile of cultured astrocyte cell membrane either after NaCl or 10 mM MSO incubation to both iMSO-Slow and iMSO-Fast mice. The more relevant lectins are shown. Data are expressed as ratio between MSO and NaCl effects standardized to PHA-E variation. Values are expressed as mean ± SEM in 5 different astrocyte cultures.

IV. Discussion

IV.1. Glycogen Changes in Cerebral Cortices

IV.1.1. Glycogen Changes in Cerebral Cortices during the Selection Process

The present data strongly suggest, in terms of seizures and brain glycogen accumulation, that our two lines of mice selected for their latency to MSO-dependent seizures (MSO-Fast and MSO-Slow) are the most extreme mouse strains actually available. The two other strains, CBA/J and C57BL/6J, are in between MSO-Fast and MSO-Slow inbred mice. In terms of latency toward MSO-dependent seizures, indeed the order of responses starting from the higher responding strain is as follows: MSO-Fast > CBA/J > C57BL/6J > MSO-Slow. Regarding glycogen variation in brain cortices induced by MSO, the order is as follows: MSO-Fast ≈ CBA/J < C57BL/6J < MSO-Slow. Therefore our set of data obtained for the ancestral generations of MSO-Fast and MSO-Slow mice strongly suggested a relationship between the two MSO-induced phenomena.

IV.1.1. Glycogen Changes in Cerebral Cortices After the Inbreeding Process

In terms of involvement of cerebral glycogen in MSO-induced seizures in MSO-Fast and MSO-Slow, i.e. after the inbreeding process of the two strains of mice, we can assume that an increase in brain glycogen content may constitute a way to deal with MSO-dependent seizures, i.e. higher brain glycogen content, higher MSO-dependent seizure latency. The hypothesis, that the ability to increase brain glycogen content may constitute a way to postpone the MSO-dependent seizures, has already been proposed according to our data obtained with CBA/J and C57BL/6J (Bernard-Hélary et al.., 2000, Bernard-Hélary et al.., 2002) and with the ancestral MSO-Fast (G6) and MSO-Slow (G6) mice during their selection process (Cloix and Hevor, 2009, Cloix et al.., 2010b). However, our new strains of iMSO-Fast and iMSO-Slow allowed us to address the role of cerebral glycogen of cortices even further and indeed our data (Figure 10) show higher glycogen levels for iMSO-Fast mice than for iMSO-Slow. Because since iMSO-Fast is still the strain with the faster response toward MSO-induced seizure activity we were able to show for the first time that glycogen by itself does not postpone seizures and may rather be used to adapt the brain in a way so that it can provide the energy needed to support seizure prone neurons in a more optimal way. With the present data, it seems that the previous hypothesis may now be abandoned. Therefore, the inbreeding process resulting in iMSO-Fast mice has led to an adaptation towards a higher energy demand in the newly generated iMSO-Fast strain when compared to iMSO-Slow, whereas the mechanisms contributing to the faster appearance of MSO-induced seizures in the iMSO-Fast strain remained almost unaltered, despite the fact that both strains increased their glycogen content in response to the convulsive dose of MSO. The transposition of this hypothesis to other models of epilepsy is not yet validated, neither is its extension to human disease. Nevertheless, quantification of brain glycogen is now considered as a putative diagnostic and therapeutic tool (Cloix et al.., 2008, Cloix and Hevor, 2009). Indeed, a great deal of effort is being made to determine brain glycogen content using non invasive methods, both in animals and humans (Gruetter, 2002, Gruetter et al.., 2003, Oz et al.., 2003, Oz et al.., 2007, van Heeswijk et al.., 2010). Moreover, as noted earlier, brain glycogen content increases in human epilepsy (Dalsgaard et al.., 2007). Tagliabracci and co-workers implicated abnormal glycogen accumulation as the cause of Lafora disease, a disorder characterized by progressive myoclonic epilepsy (Delgado-Escueta, 2007, Tagliabracci et al.., 2008). These reports are in agreement with the efforts made to determine brain glycogen content in humans, and with the proposed hypothesis of involvement of brain glycogen content in either the control of epilepsy or as an adaptation to higher energy demand within the epileptic brain as may be concluded according to results obtained for iMSO-Fast mice current study.

Taking our new data into account they could be quite consistent with the reported elevation in human brain glycogen in epileptic patients (Dalsgaard et al.., 2007) because also the long lasting time of ongoing seizures may lead to an additional, plastic increase in brain glycogen content in perifocal cerebral tissue in order to better support neurons which have to cope with an increased risk of hyperexcitation.

However, since in both strains iMSO-Fast and iMSO-Slow cerebral glycogen increased which was more pronounced in iMSO-Fast it may be possible that the iMSO-Slow strain does not only increase cerebral glycogen but at the same uses it more efficiently during the preconvulsive period which may contribute to a delay in seizure appearance. Therefore, future experiments have to be focused on developing techniques which will overcome currently existing methodological constraints to address this question in a reliable manner.

IV.2. Glycosylation Profiles

This high level of brain glycogen content in MSO-Fast inbred mice could contribute to the observed differences between MSO-Fast and MSO-Slow inbred mice in terms of glycosylation profile. Cell surface glycosylation can indeed be modulated or modified after injection of a pharmacological compound and lectins are a tool to observe these modifications (Braza-Boils et al.., 2006). In cultured astrocytes from iMSO-Slow inbred mice, MSO induces an elevation of the proportion of glycans while a diminution is observed for astrocytes derived from iMSO-Fast. First of all these in vitro data provided clear evidence that the astrocytes of both strains responded differently to MSO under otherwise similar conditions. The phenotypic difference in the glycosylation response between the two inbred lines indicated that the resistance to MSO of iMSO-Slow and the sensitivity of iMSO-Fast involves two different mechanisms in terms of glycosylation. These data would strongly support an involvement of the extracellular matrix which relates to the glia-neuron interface and its discussed role in epilepsy including also G protein-coupled receptors, whose expression is mostly regulated by N-glycans (Lanctot et al.., 2006) and/or changes related to extracellular matrix controlled perisynaptic neurotransmitter diffusion affecting perisynaptic receptors (Vargova and Sykova, 2008, Zamecnik et al.., 2012).

IV.3. Role of Neurotransmitters in Cerebral Cortices

The implication of neurotransmitters in the control of glycogen level has been largely documented and demonstrated; particularly, for 5-HT and brain glycogen content (Leonard, 1975, Magistretti, 1988, Darvesh and Gudelsky, 2003, Verleysdonk et al.., 2005). Moreover, hypofunctional serotonergic neurotransmission was described in Lafora epilepsy, and 5-hydroxytryptophan administration was found to be beneficial for patients (Pranzatelli et al.., 1995). Reduced levels of serotonergic receptors have been reported in temporal epileptic foci (Liew et al.., 2009), and 5-HT could play an important role in modulating seizure activity because it appears to inhibit the expression of overt seizure behaviors (Trindade-Filho et al.., 2008). Furthermore, the involvement of 5-HT in the control of MSO-dependent seizures has been suggested (Blizard and Balkoski, 1982); and the present data indicate that 5-HT might be involved in the response toward MSO-induced seizures in all tested strains. The basal level of 5-HT is indeed lower in iMSO-Slow mice compared to iMSO-Fast ones. This might be particularly true considering the potential of 5-HT to antagonize seizures, and the high utilization of 5-HT by synapses to respond to the changes brought about by seizures. In view of the literature discussed above it was expected that iMSO-Fast mice having such high 5-HT levels would be better equipped to suppress seizures compared to iMSO-Slow because the latter seem to be 5-HT deficient. However, like for the MSO-induced glycogen response (see above) our results show highest 5-HT in the strain with the shortest latency. Therefore, it could be assumed that the mice most prone to excitation may have adapted plastic cerebral changes towards mechanisms to increase 5-HT faster in order to cope more efficiently with the sequelae brought about by hyperexcitation. Whereas the adaptational pressure of the seizure resistant strain iMSO-Slow remained lower and accordingly 5-HT responses toward a convulsive dose of MSO remained low. The latter assumption may be also supported also by the observation that in iMSO-Fast mice not only the MSO-induced increase was affected but also basal 5-HT level was significantly higher compared to iMSO-Slow (see Fig. 7). Otherwise one would have to assume that iMSO-Slow mice use and catabolise their synthesized 5-HTP and 5-HT much faster than iMSO-Fast in order to delay seizures during their prolonged latency period. Our results need further experimental evaluations in order to find out whether the MSO challenge during the selection process may indeed have altered 5-HT level and 5-HT - utilization in brains of selected mice.

IV.4. Possible Involvement of Glutamatergic Pathway in MSO-Induced Seizures

The better sensitivity of the MSO-Fast line to the analogue of glutamate, KA, as compared to the GABA antagonist, PTZ, supposes an involvement of glutamatergic pathways in the action of MSO. Therefore our data indicate that MSO-induced seizure activity acts at least partly through a similar mechanism as KA and those mice which are better adapted to cope with MSO-induced epileptic activity such as iMSO-Slow are using the same mechanism to reduce the effects caused by KA. Therefore one part of the explanation why MSO-Fast mice are more prone to seizures than MSO-Slow relates to their difference in sensitivity towards glutamate. The analyses of KA-dependent latency showed that MSO-Slow mice are more resistant to KA-induced seizures than MSO-Fast ones, which is in agreement with a lower involvement of glutamatergic pathways in MSO-Slow lines. Such an idea is reinforced by the anticonvulsant effect of the MK-801, which blocked selectively NMDA glutamatergic receptors, while less anticonvulsant action was observed with VPA. The present data strongly suggest less involvement of GABAergic pathways as compared to glutamatergic ones in the mechanisms of MSO-dependent seizures. This observation is corroborated by the higher level of GS activity in MSO-Slow compared to the MSO-Fast line, with no significant variation in terms of affinity for glutamate and constant inhibition by MSO. These data indicate a higher conversion of glutamate into glutamine leading to a putative, yet to be demonstrated, lower level of glutamate in MSO-Slow line. The fate of brain glutamate under the influence of MSO, which was utilized to select the mice, is unclear. Nevertheless, the powerful inhibition of GS by MSO was thought to trigger the seizures, as glutamate was supposed to increase as a consequence of GS inhibition. However until today this view is heavily debated, since a decrease in brain glutamate level was observed after the administration of convulsive doses of MSO (Engelsen and Fonnum, 1985, Fonnum and Paulsen, 1990) whereas no variation had been described by Somers and Beckstead (1990). Still an unilateral deficiency of hippocampal GS, induced by MSO, causes seizures in rats (Eid et al.., 2008). In addition, in rats with PTZ-induced repetitive epileptic seizures the regions showing a strong glial heat shock response correspond with reduced GS-activity and GS-nitration, which together are clear indicators of a nitrosative stress response (Bidmon et al.., 2008) and a recent study shows MSO-induced glutamate accumulation in hippocampal astrocytes (Perez et al.., 2012). As GS is confined to astrocytes, Phelps (Phelps, 1975) suggested that glutamate might be massively drawn on to build the glycogen molecule, since glycogen accumulates to tremendous levels in the brain under the effect of MSO (Folbergrova et al.., 1969, Hevor, 1994, Cloix and Hevor, 2009, Cloix et al.., 2010a). For these reasons, the actual role of the glutamatergic system in the epileptogeny generated by MSO remains unclear and needs further investigations. In addition and in contrast to PTZ-induced seizure (Cremer et al.., 2009) no significant changes in ligand binding towards glutamate receptors of $GABA_A$ receptors had been observed in MSO challenged rats which showed only slight significant alterations for adenosine receptors within hippocampus and amygdala (Cremer et al.., 2010). Taking into account the data discussed above such as minor alterations in cerebral glutamate content and a lack of adaptive receptor alteration, one possible explanation could be that GS the enzyme responsible for glutamate catabolism is solely confined to astrocytes. As already suggested for nitrotyrosine-inhibted GS (Bidmon et al.., 2008) GS inhibition caused by MSO could lead to a pathological microcompartimentalization of glutamate in time without altering the total glutamate concentration within the affected tissue, a view which is supported by the MSO-induced unilateral glutamate accumulation reported by Perez et al.. (2012). The resulting dysbalance in glutamate distribution could cause epileptic activity either by an episodic excessive glutamate overflow from astrocytes or by other changes which may be related to astrocytic glutamate accumulation such as glycogen accumulation and altered protein glycosylation as discussed above.

IV.5. Behavioral Changes

The relationship between brain glutamatergic pathways and epilepsy has already been reported for human epilepsy as well as for animal models of epilepsy (Armijo et al.., 2005, Andrade and Minassian, 2007, Sierra-Paredes and Sierra-Marcuno, 2007, Vincent and Mulle, 2009). Mood and psychiatric disorders in patients have been described as consequences of *status epilepticus* (Kanner, 2008, 2009, Mula and Monaco, 2009). In addition, the population of epileptic patients is more prone to commit suicide (Kanner, 2009); but whether mood and/or psychiatric disorders are related to seizure genesis, or the inverse, remains an unsolved question (Kanner, 2008, 2009). In our experiments the behavior of the two inbred lines of mice was different and indicate that iMSO-Fast mice were more anxious than MSO-Slow. Many reports underline the involvement of the glutamatergic system in learning capacity and in anxiety (Mohler, 2006, Hashimoto et al.., 2007, Venault and Chapouthier, 2007b, Kanner, 2009, Duvoisin et al.., 2010a, Duvoisin et al.., 2010b). In addition, receptors of AMPA and kainate types have been reported to be implicated in memory processes (Perkinton et al.., 1999, Zilles et al.., 2000), while NMDA and metabotropic glutamatergic receptors (mGluR) have been described as involved in anxiety (Nowak et al.., 1995, Platenik et al.., 2000, Duvoisin et al.., 2010a, Duvoisin et al.., 2010b). Moreover, one example is the direct binding of glutamate to cell membranes: glutamate binding to hippocampal membrane fractions of fully kindled rats was significantly higher when compared to controls, the fully kindled rats having poorer learning performance, whereas binding of not fully kindled ones did not differ from that of controls (Rossler et al.., 2000). In our experiments, the difference in GS activity may account for the difference in anxiety behavior between the two lines due to the pivotal role of the latter enzyme in the glutamate/glutamine pathway. Indeed, even though GS is confined to astrocytes, the latter provide glutamine to neurons, which transform it into glutamate and GABA. Astrocytes withdraw glutamate released by glutamatergic neurons in the synaptic cleft; thus the magnitude of GS activity may influence the glutamate/glutamine cycle in the brain. As this enzyme activity is statistically different between the two inbred strains of mice, it is highly probable that this impacts the behavior of each line differently. In future experiments, blocking glutamate receptors may help to clarify the relationship between the glutamatergic system, behavior, and seizure genesis. Moreover, we reported a higher level of serotonin in the cortex of MSO-Fast compared to MSO-Slow mice (Cloix et al.., 2010a) which could also contribute to the anxiety developed by MSO-Fast mice. This might fit with what is known in depressed patients that have elevated expression of neuronal tryptophan hydroxylase 2 mRNA and protein in midbrain serotonergic neurons, as well as increases in brain serotonin turnover (Lowry et al.., 2008) indicating that both our observed alterations in the amount of 5-HTP and 5-HT might contribute to the observed behavioral differences.

The involvement of the glutamatergic system in the mechanism of MSO epileptogenic activity had been strongly suggested years ago (Griffith and Meister, 1978); but more recent data point to the prominent role of mGluR in various psychiatric disorders in human and in models of mood disorders (Wieronska and Pilc, 2009, Duvoisin et al.., 2010a, Duvoisin et al.., 2010b, Gravius et al.., 2010, Krystal et al.., 2010). Although the role of glutamate was disclaimed in the epileptogenic action of MSO, because of the lack of increase in glutamate content (Engelsen and Fonnum, 1985, Fonnum and Paulsen, 1990, Somers and Beckstead, 1990), recent data (Berlicki, 2008, Eid et al.., 2008, Zou et al.., 2010), together with the present results, this strongly suggest the necessity of a revalidation of the glutamate pathways in the mechanism of action of MSO.

The design of the investigation used during the selection process of iMSO-Fast and iMSO-Slow mice has been described in another model of epilepsy based on the GABAergic system. β-Carboline, an inverse agonist of GABAergic receptors, was used to select two inbred lines of mice having different sensitivity toward seizure genesis (Chapouthier et al.., 1998). Data obtained with this mouse model suggested that the two strains differed in their motor activity, anxiety, aggression, and learning capacity (Guillot et al.., 1999, Rossler et al.., 2000, Suaudeau et al.., 2000, Venault et al.., 2006, Venault and Chapouthier, 2007a). Previous data, in addition to the present work, underlines a parallelism between an impairment of behavior and seizure genesis. The basis of the actual relationship between these two physiological effects remains to be established.

Conclusion

The present investigation compares some characteristics of two commercially available inbred strains of mice, CBA/J and C57BL/6J, with two selected strains of mice, iMSO-Fast and iMSO-Slow, according to their latency toward MSO-dependent seizures. The main result describes a relationship between seizures latency and the accumulation of brain glycogen, the iMSO-Fast and iMSO-Slow mice being the more extreme. This allowed us for the first time to address the role of glycogen accumulation related to epileptic activity in a different way since in our inbred animal model higher glycogen accumulation is associated with a faster and more severe seizure response which contradicts the current discussion about the functional relation between glycogen and epileptic activity. The implication of brain 5-HT in the responses to MSO, both in terms of seizures and glycogen, seems to be relevant. In addition, iMSO-Fast mice seem more anxious than the MSO-Slow ones. We believe that the use of inbred mice, and particularly the mice newly selected for their sensitivity toward MSO, may be very helpful to better understand the mechanisms underlying the responses to MSO, and could also be useful to extrapolate data obtained with animal models to human disease.

Acknowledgment

This work was supported by the Ministère de l'Education Nationale, de la Recherche et de l'Enseignement Supérieur, by the Centre National de la Recherche Scientifique (C.N.R.S.), by the Région Centre, and the Conseil Général du Loiret. The work is in partial fulfillment of the requirements for the degree of Ph.D. at the University of Orléans (A.B.). We are grateful to Ph Moreau for his helpful technical assistance. Authors disclose no conflict of interests.

References

(CCTILAE), 1989. Proposal for revised classification of epilepsies and epileptic syndromes. Commission on Classification and Terminology of the International League Against Epilepsy. *Epilepsia* 30, 389-399.

Andrade, DM, Minassian, BA, 2007. Genetics of epilepsies. *Expert Rev Neurother* 7, 727-734.

Armijo, JA, Shushtarian, M, Valdizan, EM, Cuadrado, A, de las Cuevas, I, Adin, J, 2005. Ion channels and epilepsy. *Curr Pharm Des* 11, 1975-2003.

Belzung, C, Griebel, G, 2001. Measuring normal and pathological anxiety-like behaviour in mice: a review. *Behav Brain Res* 125, 141-149.

Berlicki, L, 2008. Inhibitors of glutamine synthetase and their potential application in medicine. *Mini Rev Med Chem* 8, 869-878.

Bernard-Hélary, K, Ardourel, M, Hévor, T, Cloix, J-F, 2002. *In vivo* and *in vitro* glycogenic effects of methionine sulfoximine are different in two inbred strains of mice. *Brain Res* 929, 147-155.

Bernard-Hélary, K, Lapouble, E, Ardourel, M, Hevor, T, Cloix, J-F, 2000. Correlation between brain glycogen and convulsive state in mice submitted to methionine sulfoximine. *Life Sci* 67, 1773-1781.

Bidmon, HJ, Gorg, B, Palomero-Gallagher, N, Schleicher, A, Haussinger, D, Speckmann, EJ, Zilles, K, 2008. Glutamine synthetase becomes nitrated and its activity is reduced during repetitive seizure activity in the pentylentetrazole model of epilepsy. *Epilepsia* 49, 1733-1748.

Blin, M, Crusio, WE, Hevor, T, Cloix, J-F, 2002. Chronic inhibition of glutamine synthetase is not associated with impairment of learning and memory in mice. *Brain Res Bull* 57, 11-15.

Blizard, DA, Balkoski, V, 1982. Tryptophan availability, central serotonergic function and methionine sulphoximine-induced convulsions. *Neuropharmacology* 21, 27-30.

Boissonnet, A, Hevor, T, Cloix, JF, 2012. Phenotypic differences between Fast and Slow methionine sulfoximine-inbred mice: seizures, anxiety, and glutamine synthetase. *Epilepsy Res* 98, 25-34.

Braza-Boils, A, Tomas, M, Marin, MP, Megias, L, Sancho-Tello, M, Fornas, E, Renau-Piqueras, J, 2006. Glycosylation is altered by ethanol in rat hippocampal cultured neurons. *Alcohol Alcohol* 41, 494-504.

Brown, AM, 2004. Brain glycogen re-awakened. *J Neurochem* 89, 537-552.

Brown, AM, Sickmann, HM, Fosgerau, K, Lund, TM, Schousboe, A, Waagepetersen, HS, Ransom, BR, 2005. Astrocyte glycogen metabolism is required for neural activity during aglycemia or intense stimulation in mouse white matter. *J Neurosci Res* 79, 74-80.

Chapouthier, G, Launay, JM, Venault, P, Breton, C, Roubertoux, PL, Crusio, WE, 1998. Genetic selection of mouse lines differing in sensitivity to a benzodiazepine receptor inverse agonist. *Brain Res* 787, 85-90.

Cloix, JF, Ardourel, MY, Hévor, T, 2008. Glycogen metabolism and brain pathologies. *Central Nervous System Agents in Medicinal Chemistry* 8, 187-197.

Cloix, JF, Hevor, T, 2009. Epilepsy, regulation of brain energy metabolism and neurotransmission. *Curr Med Chem* 16, 841-853.

Cloix, JF, Tahi, Z, Boissonnet, A, Hevor, T, 2010a. Brain glycogen and neurotransmitter levels in fast and slow methionine sulfoximine-selected mice. *Exp Neurol* 225, 38-47.

Cloix, JF, Tahi, Z, Martin, B, Hevor, T, 2010b. Selection of two lines of mice based on latency to onset of methionine sulfoximine seizures. *Epilepsia* 51, 118-128.

Cremer, CM, Bidmon, HJ, Gorg, B, Palomero-Gallagher, N, Escobar, JL, Speckmann, EJ, Zilles, K, 2010. Inhibition of glutamate/glutamine cycle in vivo results in decreased benzodiazepine binding and differentially regulated GABAergic subunit expression in the rat brain. *Epilepsia*

Cremer, CM, Palomero-Gallagher, N, Bidmon, HJ, Schleicher, A, Speckmann, EJ, Zilles, K, 2009. Pentylenetetrazole-induced seizures affect binding site densities for GABA, glutamate and adenosine receptors in the rat brain. *Neuroscience* 163, 490-499.

Crusio, WE, 2001. Genetic dissection of mouse exploratory behaviour. *Behav Brain Res* 125, 127-132.

Dalsgaard, MK, Madsen, FF, Secher, NH, Laursen, H, Quistorff, B, 2007. High glycogen levels in the hippocampus of patients with epilepsy. *J Cereb Blood Flow Metab* 27, 1137-1141.

Darvesh, AS, Gudelsky, GA, 2003. Activation of 5-HT2 receptors induces glycogenolysis in the rat brain. *Eur J Pharmacol* 464, 135-140.

Delgado-Escueta, AV, 2007. Advances in lafora progressive myoclonus epilepsy. *Curr Neurol Neurosci Rep* 7, 428-433.

Dreifuss, FE, 1987. Classification of the epilepsies: influence on management. *Rev Neurol* (Paris) 143, 375-380.

Dudek, FE, Patrylo, PR, Wuarin, JP, 1999. Mechanisms of neuronal synchronization during epileptiform activity. *Adv Neurol* 79, 699-708.

Duvoisin, RM, Pfankuch, T, Wilson, JM, Grabell, J, Chhajlani, V, Brown, DG, Johnson, E, Raber, J, 2010a. Acute pharmacological modulation of mGluR8 reduces measures of anxiety. *Behav Brain Res* 212, 168-173.

Duvoisin, RM, Villasana, L, Pfankuch, T, Raber, J, 2010b. Sex-dependent cognitive phenotype of mice lacking mGluR8. *Behav Brain Res* 209, 21-26.

Eid, T, Ghosh, A, Wang, Y, Beckstrom, H, Zaveri, HP, Lee, TS, Lai, JC, Malthankar-Phatak, GH, de Lanerolle, NC, 2008. Recurrent seizures and brain pathology after inhibition of glutamine synthetase in the hippocampus in rats. *Brain* 131, 2061-2070.

Eid, T, Thomas, MJ, Spencer, DD, Runden-Pran, E, Lai, JC, Malthankar, GV, Kim, JH, Danbolt, NC, Ottersen, OP, de Lanerolle, NC, 2004. Loss of glutamine synthetase in the human epileptogenic hippocampus: possible mechanism for raised extracellular glutamate in mesial temporal lobe epilepsy. *Lancet* 363, 28-37.

Einat, H, Manji, HK, Gould, TD, Du, J, Chen, G, 2003. Possible involvement of the ERK signaling cascade in bipolar disorder: behavioral leads from the study of mutant mice. *Drug News Perspect* 16, 453-463.

Engel, J, Jr., Brown, WJ, Kuhl, DE, Phelps, ME, Mazziotta, JC, Crandall, PH, 1982. Pathological findings underlying focal temporal lobe hypometabolism in partial epilepsy. *Ann Neurol* 12, 518-528.

Engel, J, Jr., Kuhl, DE, Phelps, ME, Rausch, R, Nuwer, M, 1983. Local cerebral metabolism during partial seizures. *Neurology* 33, 400-413.

Engelsen, B, Fonnum, F, 1985. The effect of methionine sulfoximine, an inhibitor of glutamine synthetase, on the levels of amino acids in the intact and decorticated rat neostriatum. *Brain Res* 338, 165-168.

Everitt, AD, Sander, JW, 1999. Classification of the epilepsies: time for a change? A critical review of the International Classification of the Epilepsies and Epileptic Syndromes (ICEES) and its usefulness in clinical practice and epidemiological studies of epilepsy. *Eur Neurol* 42, 1-10.

Folbergrova, J, 1973. Glycogen and glycogen phosphorylase in the cerebral cortex of mice under the influence of methionine sulphoximine. *J Neurochem* 20, 547-557.

Folbergrova, J, Druga, R, Haugvicova, R, Mares, P, Otahal, J, 2008. Anticonvulsant and neuroprotective effect of (S)-3,4-dicarboxyphenylglycine against seizures induced in immature rats by homocysteic acid. *Neuropharmacology* 54, 665-675.

Folbergrova, J, Passonneau, JV, Lowry, OH, Schulz, DW, 1969. Glycogen, ammonia and related metabolites in the brain during seizures evoked by methionine sulphoximine. *J Neurochem* 16, 191-203.

Fonnum, F, 1984. Glutamate: A neurotransmitter in mammalian brain. *J Neurochem* 42, 1-11.

Fonnum, F, Paulsen, RE, 1990. Comparison of transmitter amino acid levels in rat globus pallidus and neostriatum during hypoglycemia or after treatment with methionine sulfoximine or gamma-vinyl gamma-aminobutyric acid. *J Neurochem* 54, 1253-1257.

Gastaut, H, Toga, M, Naquet, R, 1958. Clinical, electrographical and anatomical study of epilepsy induced in dogs by ingestion of agenised proteins. In *Temporal Lobe Epilepsy;* Ed; Baldwin, M; Bailey P, Thomas Publisher, Illinois 57-69.

Gravius, A, Pietraszek, M, Dekundy, A, Danysz, W, 2010. Metabotropic glutamate receptors as therapeutic targets for cognitive disorders. *Curr Top Med Chem* 10, 187-206.

Griffith, OW, Meister, A, 1978. Differential inhibition of glutamine and gamma-glutamylcysteine synthetases by alpha-alkyl analogs of methionine sulfoximine that induce convulsions. *J Biol Chem* 253, 2333-2338.

Gruetter, R, 2002. In vivo 13C NMR studies of compartmentalized cerebral carbohydrate metabolism. *Neurochem Int* 41, 143-154.

Gruetter, R, Adriany, G, Choi, IY, Henry, PG, Lei, H, Oz, G, 2003. Localized in vivo 13C NMR spectroscopy of the brain. *NMR Biomed* 16, 313-338.

Guillot, PV, Sluyter, F, Crusio, WE, Chapouthier, G, 1999. Mice selected for differences in sensitivity to a benzodiazepine receptor inverse agonist vary in intermale aggression. *Neurogenetics* 2, 171-175.

Hashimoto, K, Sawa, A, Iyo, M, 2007. Increased levels of glutamate in brains from patients with mood disorders. *Biol Psychiatry* 62, 1310-1316.

Hevor, TK, 1994. Some aspects of carbohydrate metabolism in the brain. *Biochimie* 76, 111-120.

Hevor, TK, Delorme, P, 1991. Biochemical and ultrastructural study of glycogen in cultured astrocytes submitted to the convulsant methionine sulfoximine. *Glia* 4, 64-69.

Hevor, TK, Delorme, P, Beauvillain, JC, 1986. Glycogen synthesis and immunocytochemical study of fructose-1,6- biphosphatase in methionine sulfoximine epileptogenic rodent brain. *J Cereb Blood Flow Metab* 6, 292-297.

Hevor, TK, Delorme, P, Gayet, J, 1985. Glycogen content and fructose-1, 6-biphosphatase activity in methionine sulfoximine epileptogenic mouse brain and liver after protein synthesis inhibition. *Neuropathol Appl Neurobiol* 11, 129-139.

Johnston, MV, 1996. Developmental aspects of epileptogenesis. *Epilepsia* 37 Suppl 1, S2-9.

Kanner, AM, 2008. The use of psychotropic drugs in epilepsy: what every neurologist should know. *Semin Neurol* 28, 379-388.

Kanner, AM, 2009. Psychiatric issues in epilepsy: the complex relation of mood, anxiety disorders, and epilepsy. *Epilepsy Behav* 15, 83-87.

King, JT, Jr., LaMotte, CC, 1989. El mouse as a model of focal epilepsy: a review. *Epilepsia* 30, 257-265.

Krystal, JH, Mathew, SJ, D'Souza, DC, Garakani, A, Gunduz-Bruce, H, Charney, DS, 2010. Potential psychiatric applications of metabotropic glutamate receptor agonists and antagonists. *CNS Drugs* 24, 669-693.

Lai, V, Mak, HK, Yung, AW, Ho, WY, Hung, KN, 2010. Neuroimaging techniques in epilepsy. *Hong Kong Med J* 16, 292-298.

Lanctot, PM, Leclerc, PC, Escher, E, Guillemette, G, Leduc, R, 2006. Role of N-glycan-dependent quality control in the cell-surface expression of the AT1 receptor. *Biochem Biophys Res Commun* 340, 395-402.

Landemarre, L, Cancellieri, P, Duverger, E, 2012. Cell surface lectin-array: parameters affecting cell glycan signature. *Glycoconjugate J* Aug 17.

Landemarre, L, Duverger, E, 2013. Lectin glycoprofiling of recombinant therapeutic Interleukin-7. *Meth Mol Biol* 988.

Leonard, BE, 1975. The effect of 5-hydroxytryptamine and histamine on glycolysis in the mouse brain. *Z Naturforsch* C 30, 113-116.

Liew, CJ, Lim, YM, Bonwetsch, R, Shamim, S, Sato, S, Reeves-Tyer, P, Herscovitch, P, Dustin, I, Bagic, A, Giovacchini, G, Theodore, WH, 2009. 18F-FCWAY and 18F-FDG PET in MRI-negative temporal lobe epilepsy. *Epilepsia* 50, 234-239.

Lowry, CA, Hale, MW, Evans, AK, Heerkens, J, Staub, DR, Gasser, PJ, Shekhar, A, 2008. Serotonergic systems, anxiety, and affective disorder: focus on the dorsomedial part of the dorsal raphe nucleus. *Ann N Y Acad Sci* 1148, 86-94.

Lowry, OH, Rosebrough, NJ, Farr, AL, Randall, RJ, 1951. Protein measurement with the Folin phenol reagent. *J Biol Chem* 193, 265-275.

Magistretti, PJ, 1988. Regulation of glycogenolysis by neurotransmitters in the central nervous system. *Diabete Metab* 14, 237-246.
Magistretti, PJ, 2006. Neuron-glia metabolic coupling and plasticity. *J Exp Biol* 209, 2304-2311.
Magistretti, PJ, Pellerin, L, 1996. The contribution of astrocytes to the ^{18}F-2-deoxyglucose signal in PET activation studies. *Mol Psych* 1, 445-452.
Magistretti, PJ, Pellerin, L, Rothman, DL, Shulman, RG, 1999. Energy on demand. *Science* 283, 496-497.
McCormick, DA, Contreras, D, 2001. On the cellular and network bases of epileptic seizures. *Annu Rev Physiol* 63, 815-846.
Meister, A, 1978. Inhibition of glutamine synthase and gamma glutamylcysteine synthetase by methionine sulfoximine and related compounds. In *Enzyme Activated Irreversible Inhibitiors;* Ed; Seiler, N; Jung, MJ; Koch-Weser, J Elsevier, Amsterdam, The Netherlands 187-210.
Mohler, H, 2006. GABAA receptors in central nervous system disease: anxiety, epilepsy, and insomnia. *J Recept Signal Transduct Res* 26, 731-740.
Mula, M, Monaco, F, 2009. Antiepileptic drugs and psychopathology of epilepsy: an update. *Epileptic Disord* 11, 1-9.
Nowak, G, Ordway, GA, Paul, IA, 1995. Alterations in the N-methyl-D-aspartate (NMDA) receptor complex in the frontal cortex of suicide victims. *Brain Res* 675, 157-164.
Oz, G, Henry, PG, Seaquist, ER, Gruetter, R, 2003. Direct, noninvasive measurement of brain glycogen metabolism in humans. *Neurochem Int* 43, 323-329.
Oz, G, Seaquist, ER, Kumar, A, Criego, AB, Benedict, LE, Rao, JP, Henry, PG, Van De Moortele, PF, Gruetter, R, 2007. Human brain glycogen content and metabolism: implications on its role in brain energy metabolism. *Am J Physiol Endocrinol Metab* 292, E946-951.
Pape, HC, Jungling, K, Seidenbecher, T, Lesting, J, Reinscheid, RK, 2010. Neuropeptide S: a transmitter system in the brain regulating fear and anxiety. *Neuropharmacology* 58, 29-34.
Pape, HC, Pare, D, 2010. Plastic synaptic networks of the amygdala for the acquisition, expression, and extinction of conditioned fear. *Physiol Rev* 90, 419-463.
Pellegri, G, Rossier, C, Magistretti, PJ, Martin, JL, 1996. Cloning, localization and induction of mouse brain glycogen synthase. *Brain Res Mol Brain Res* 38, 191-199.
Perez, EL, Lauritzen, F, Wang, Y, Lee, TS, Kang, D, Zaveri, HP, Chaudhry, FA, Ottersen, OP, Bergersen, LH, Eid, T, 2012. Evidence for astrocytes as a potential source of the glutamate excess in temporal lobe epilepsy. *Neurobiology of Disease* 10.1016/j.nbd.2012.05.010,
Perkinton, MS, Sihra, TS, Williams, RJ, 1999. Ca(2+)-permeable AMPA receptors induce phosphorylation of cAMP response element-binding protein through a phosphatidylinositol 3-kinase-dependent stimulation of the mitogen-activated protein kinase signaling cascade in neurons. *J Neurosci 1*9, 5861-5874.
Peters, EL, Tower, DB, 1959. Glutamic acid and glutamine metabolism in cerebral cortex after seizures induced by methionine sulphoximine. *J Neurochem* 5, 80-90.
Phelps, CH, 1975. An ultrastructural study of methionine sulphoximine-induced glycogen accumulation in astrocytes of the mouse cerebral cortex. *J Neurocytol* 4, 479-490.

Platenik, J, Kuramoto, N, Yoneda, Y, 2000. Molecular mechanisms associated with long-term consolidation of the NMDA signals. Life Sci 67, 335-364.

Pranzatelli, MR, Nadi, NS, 1995. Mechanism of action of antiepileptic and antimyoclonic drugs. *Adv Neurol* 67, 329-360.

Pranzatelli, MR, Tate, E, Huang, Y, Haas, RH, Bodensteiner, J, Ashwal, S, Franz, D, 1995. Neuropharmacology of progressive myoclonus epilepsy: response to 5-hydroxy-L-tryptophan. *Epilepsia* 36, 783-791.

Rampp, S, Stefan, H, 2006. Fast activity as a surrogate marker of epileptic network function? *Clin Neurophysiol* 117, 2111-2117.

Rossler, AS, Schroder, H, Dodd, RH, Chapouthier, G, Grecksch, G, 2000. Benzodiazepine receptor inverse agonist-induced kindling of rats alters learning and glutamate binding. *Pharmacol Biochem Behav* 67, 169-175.

Scharfman, HE, 2007. The neurobiology of epilepsy. *Curr Neurol Neurosci Rep* 7, 348-354.

Sierra-Paredes, G, Sierra-Marcuno, G, 2007. Extrasynaptic GABA and glutamate receptors in epilepsy. *CNS Neurol Disord Drug Targets* 6, 288-300.

Sitoh, YY, Tien, RD, 1998. Neuroimaging in epilepsy. *J Magn Reson Imaging* 8, 277-288.

Somers, DL, Beckstead, RM, 1990. Chronic methionine sulfoximine administration reduces synaptosomal aspartate and glutamate in rat striatum. *Neurosci Lett* 115, 335-340.

Suaudeau, C, Rinaldi, D, Lepicard, E, Venault, P, Crusio, WE, Costentin, J, Chapouthier, G, 2000. Divergent levels of anxiety in mice selected for differences in sensitivity to a convulsant agent. *Physiol Behav* 71, 517-523.

Tagliabracci, VS, Girard, JM, Segvich, D, Meyer, C, Turnbull, J, Zhao, X, Minassian, BA, Depaoli-Roach, AA, Roach, PJ, 2008. Abnormal metabolism of glycogen phosphate as a cause for Lafora disease. *J Biol Chem* 283, 33816-33825.

Trindade-Filho, EM, de Castro-Neto, EF, de, ACR, Lima, E, Scorza, FA, Amado, D, Naffah-Mazzacoratti Mda, G, Cavalheiro, EA, 2008. Serotonin depletion effects on the pilocarpine model of epilepsy. *Epilepsy Res* 82, 194-199.

Uhlhaas, PJ, Singer, W, 2006. Neural synchrony in brain disorders: relevance for cognitive dysfunctions and pathophysiology. *Neuron* 52, 155-168.

van Heeswijk, RB, Morgenthaler, FD, Xin, L, Gruetter, R, 2010. Quantification of brain glycogen concentration and turnover through localized 13C NMR of both the C1 and C6 resonances. *NMR Biomed* 23, 270-276.

Vargova, L, Sykova, E, 2008. Extracellular space diffusion and extrasynaptic transmission. *Physiol Res* 57 Suppl 3, S89-99.

Venault, P, Beracochea, D, Valleau, M, Joubert, C, Chapouthier, G, 2006. Mouse lines selected for difference in sensitivity to beta-CCM also differ in memory processes. *Behav Brain Res* 173, 282-287.

Venault, P, Chapouthier, G, 2007a. From the behavioral pharmacology of beta-carbolines to seizures, anxiety, and memory. *ScientificWorldJournal* 7, 204-223.

Venault, P, Chapouthier, G, 2007b. Plasticity and anxiety. *Neural Plast* 2007, 75617.

Verge, V, Legrand, A, Hevor, T, 1996. Isolation of carbohydrate metabolic clones from cultured astrocytes. *Glia* 18, 244-254.

Verleysdonk, S, Kistner, S, Pfeiffer-Guglielmi, B, Wellard, J, Lupescu, A, Laske, J, Lang, F, Rapp, M, Hamprecht, B, 2005. Glycogen metabolism in rat ependymal primary cultures: regulation by serotonin. *Brain Res* 1060, 89-99.

Vilchez, D, Ros, S, Cifuentes, D, Pujadas, L, Valles, J, Garcia-Fojeda, B, Criado-Garcia, O, Fernandez-Sanchez, E, Medrano-Fernandez, I, Dominguez, J, Garcia-Rocha, M, Soriano, E, Rodriguez de Cordoba, S, Guinovart, JJ, 2007. Mechanism suppressing glycogen synthesis in neurons and its demise in progressive myoclonus epilepsy. *Nat Neurosci* 10, 1407-1413.

Vincent, P, Mulle, C, 2009. Kainate receptors in epilepsy and excitotoxicity. *Neuroscience* 158, 309-323.

Wada, JA, Ikeda, H, Berry, K, 1967. Reversible behavioral and electrographic manifestations induced by methionine sulfoximine. *Neurology* 17, 854-868.

Wang, W, Lohi, H, Skurat, AV, DePaoli-Roach, AA, Minassian, BA, Roach, PJ, 2007. Glycogen metabolism in tissues from a mouse model of Lafora disease. *Arch Biochem Biophys* 457, 264-269.

Wieronska, JM, Pilc, A, 2009. Metabotropic glutamate receptors in the tripartite synapse as a target for new psychotropic drugs. *Neurochem Int* 55, 85-97.

Wieser, HG, Blume, WT, Fish, D, Goldensohn, E, Hufnagel, A, King, D, Sperling, MR, Luders, H, Pedley, TA, 2001. ILAE Commission Report. Proposal for a new classification of outcome with respect to epileptic seizures following epilepsy surgery. *Epilepsia* 42, 282-286.

Wolfe, LS, Elliot, KAC, 1962. Chemical studies in regulation to convulsive conditions. In Ed; Elliot, KAC: Page, IH; Quastel, JH; *Neurochemistry 2nd edition,* Springfield, USA, 694-697.

Zamecnik, J, Homola, A, Cicanic, M, Kuncova, K, Marusic, P, Krsek, P, Sykova, E, Vargova, L, 2012. The extracellular matrix and diffusion barriers in focal cortical dysplasias. *Eur J Neurosci* 36, 2017-2024.

Zilles, K, Wu, J, Crusio, WE, Schwegler, H, 2000. Water maze and radial maze learning and the density of binding sites of glutamate, GABA, and serotonin receptors in the hippocampus of inbred mouse strains. *Hippocampus* 10, 213-225.

Zou, J, Wang, YX, Dou, FF, Lu, HZ, Ma, ZW, Lu, PH, Xu, XM, 2010. Glutamine synthetase down-regulation reduces astrocyte protection against glutamate excitotoxicity to neurons. *Neurochem Int* 56, 577-584.

In: Methionine
Editor: Alexander Snegursky

ISBN: 978-1-62618-246-2

Chapter IV

Methionine Metabolism in Plants: Current Understanding of the Factors Regulating Its Metabolism

Rachel Amir[*] and Yael Hacahm
Laboratory of Plant Science, MIGAL,
Galilee Technology Center, Kiryat Shmona, Israel
Tel-Hai College, Upper Galilee, Israel

Abstract

In addition to the importance of methionine as a building block of protein synthesis in plants and monogastric animals including humans; methionine plays a significant role in plant metabolism. Through its metabolite, *S*-adenosyl-methionine (SAM), methionine controls the levels of several key metabolites such as ethylene, polyamines and biotin. SAM is also the primary methyl group donor that regulates different processes in plants such as cell wall and chlorophyll synthesis. The level of methionine is tightly regulated in plants most probably because methionine associated metabolites control plant metabolism and growth. Using genetic and molecular biology tools, recent studies reveal factors and genes that are involved in methionine metabolism. Manipulation of these genes leads to an increase or decrease in methionine content in plants. This review summarizes the current knowledge and recent progress made in our understanding of the factors and enzymes that regulate methionine synthesis and its degradation in vegetative and seed tissues. The accumulated knowledge enables researchers to tailor-make improvements to methionine content in plants causing only minimal interference to plant growth and productivity. Such manipulations have already resulted in significant improvement to the nutritional value of several crop plants.

* corresponding author: Laboratory of Plant Science, MIGAL, Galilee Technology Center, P.O. Box 831, Kiryat Shmona, 11016, Israel, Tel: 972-4-6953516; Fax: 972-4-6944980; E-mail: Rachel@migal.org.il

Methionine in Plant Metabolism

Methionine, a sulfur containing amino acid plays a major role in plants metabolism. Apart from its role as a protein constituent and its central role in the initiation of mRNA translation, methionine indirectly regulates a variety of cellular processes through its precursor *S*-adenosyl methionine (SAM) (Fig. 1 and 2). In higher plants, SAM is the precursor for the hormone ethylene (Fig. 2), which regulates developmental stages including ripening and senescence [1, 2], and is the source of the propylamino group in the synthesis of the polyamines spermidine and spermine. Polyamines play crucial roles in many aspects of plant growth, including cell proliferation and differentiation, apoptosis, homeostasis, gene expression and protection against environmental stresses [3]. SAM is also the precursor for the metal ion chelating compounds nicotinamide and phytosiderophores, as well as for the biotin co-factor (vitamin B_7) [4-8]. SAM is the primary biological methyl group donor (Fig. 2) required for different SAM-dependent methyltransferases participating in the synthesis of both primary and secondary metabolites. SAM-dependent methyltransferases are involved in the synthesis of lipids, DNA, RNA, proteins, pectin, alkaloids, phytosterols and osmoprotectants. They also participate in reactions required for lignins, suberins, chlorophyll

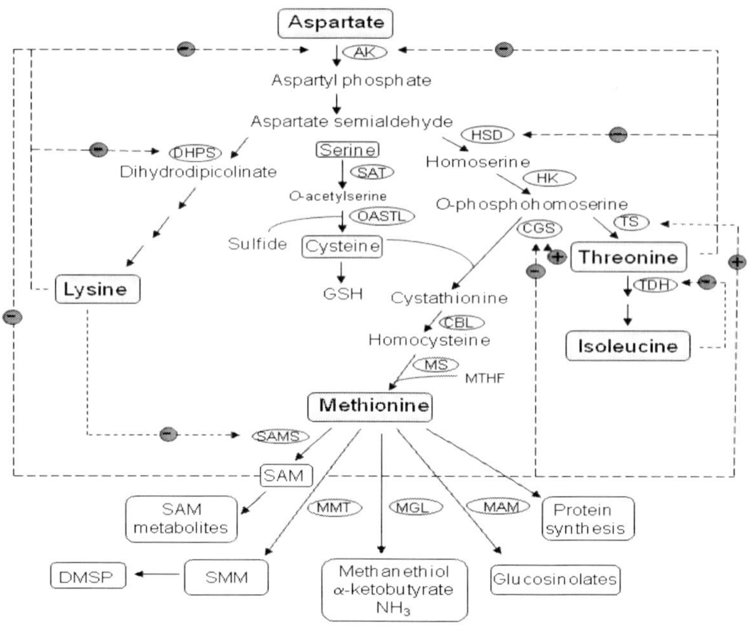

Figure 1. The aspartate family of amino acids and the biosynthesis pathways leading to methionine and cysteine in higher plants. Only some of the enzymes and metabolites are specified. Dashed arrows with a "minus" sign represent either feedback inhibition loops or repression of gene expression. The dashed and dotted arrow with the "plus" sign represents the stimulation of gene expression or enzyme activity. Abbreviations: AK, Asp kinase; DHPS, dihydrodipicolinate synthase; HSD, homoserine dehydrogenase; HK, homoserine kinase; TS, threonine synthase; TDH, threonine dehydratase;; CGS, cystathionine γ-synthase (marked in red); CBL, cystathionine β-lyase; MS, methionine synthase; SAT, serine acetyltransferase;; OASTL, *O*-acetylserine (thiol)lyase; GSH, glutathione; SAM, S-adenosyl Met; SAMS, SAM synthase; SMM, S-methylMet; MMT, methionine *S*-methyltransferase; DMSP, Dimethylsulfoniopropionate; MGL, Methionine gamma-lyase, MAM, methylthioalkylmalate synthases.

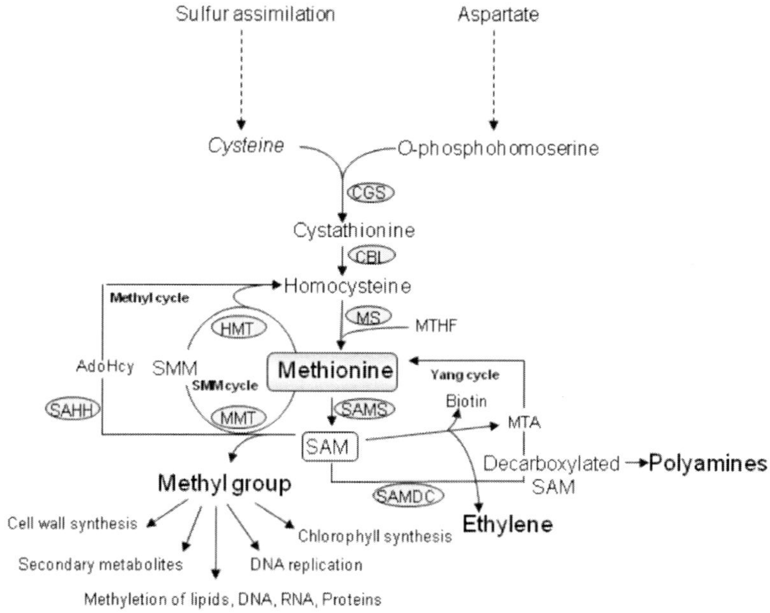

Figure 2. The three recycling pathways of methionine. Abbreviations: CGS, cystathionine γ-synthase; CBL, cystathionine β-lyase; MS, methionine synthase; SAMS, SAM synthase; SAM, *S*-adenosyl Met; SAMDC, SAM decarboxylase; MTRK, methylthioribose kinase; MTA, methylthioribose; AdoHcys, adenosylhomocysteine; SAHH, SAH hydrolase; SMM, *S*-methylMet; MMT, methionine *S*-methyltransferase; HMT, homocysteine *S*-methyltransferase.

synthesis and in the synthesis of aromatic and volatile fragrance compounds [8, 9]. Thus, methionine regulates through SAM essential cellular processes such as cell division, cell wall synthesis, chlorophyll synthesis and membrane synthesis [8].

In addition to the synthesis of SAM, methionine serves as a precursor for other metabolites. In Brassicaceae, methionine itself serves as a donor for various secondary metabolites belonging to glucosinolates (Fig. 1), which are involved in pathogen and insect defense [10, 11]. Finally, methionine leads to the synthesis of *S*-methylmethionine (SMM) (Fig. 1 and Fig. 2). SMM is considered to be the mobile and storage form of methionine [12], the regulator of SAM level in cells [13, 14], and a precursor to other secondary metabolites such as 3-Dimethyl-sulfoniopropionate, an osmoprotectant accumulated by certain flowering plants and algae [15]. 3-Dimethyl-sulfoniopropionate is also the main source of atmospheric sulfur, dimethylsulphide that has the potential to create new aerosols, which act as cloud condensation nuclei [16]. Through this interaction with cloud formation, the massive production of atmospheric dimethylsulphide over the oceans performed mainly by algae may have a significant impact on the Earth's climate.

Due to the importance of methionine metabolism in plants, methionine synthesis, accumulation and consumption are under stringent regulatory control. Considerable efforts were therefore invested to study the factors regulating its metabolism in plants, particularly in vegetative tissues (reviewed by [17-19]) and to a lesser extend in seeds. Several reviews describing different aspects of methionine metabolism were published in recent years [4, 6, 20]. This review describes the current knowledge in three sub-sections: First, it describes the regulatory role of enzymes in the methionine biosynthesis pathway. Second, the regulatory role of the methionine recycling pathways and the catabolic enzymes of methionine on its content will be described, and third, the importance of methionine as an essential amino acid for human nutrition.

The Methionine Biosynthesis Pathway in Plants

Methionine is synthesized in higher plants from a combination of three different moieties, derived from three separate pathways (Fig. 1). (i) The carbon-amino skeleton is derived from aspartate, and hence methionine is a member of the aspartate-family of amino acids, together with threonine, lysine and isoleucine (Fig. 1) [21, 22]. (ii) The sulfur moiety of methionine is derived from cysteine, the first organic metabolite of the sulfate assimilation pathway; and (iii) The methyl group of methionine is derived from *N*-methyltetrahydrofolate (Fig. 1). Cystathionine γ-synthase (CGS), the first committed enzyme of methionine biosynthesis, catalyzes a central step in methionine synthesis by joining metabolites from two separate pathways, *O*-phosphohomoserine (from the aspartate family pathway) and cysteine (from the cysteine biosynthesis pathway), to form cystathionine [23]. This compound is then converted to serine and homocysteine by cystathionine β-lyase. The effect of these two reactions is the net transfer of the thiol group of cysteine, a 3-carbon compound, to form homocysteine, a 4-carbon compound. The last enzyme in methionine biosynthesis, methionine synthase, catalyzes the methylation of homocysteine to form methionine (Fig. 1) [24].

1. Regulation of Methionine Biosynthesis

The Regulatory Role of CGS

As the first committed enzyme of methionine synthesis in plants, CGS plays a major role in the regulation of methionine synthesis. Unlike other enzymes that belong to the aspartate biosynthesis pathways, CGS activity is not controlled by feedback-inhibition mechanism [25]. Analysis of the effect of different metabolites from methionine biosynthesis and catabolic pathways did not reveal any inhibition of CGS activity [24]. The mechanism regulating *Arabidopsis* CGS (AtCGS) was discovered later by the analysis of Arabidopsis mutants that over-accumulate methionine [26]. These studies revealed that the transcript level of AtCGS is negatively regulated by methionine's downstream product, SAM [27-31]. This regulation operates via a post-transcriptional mechanism. When the level of methionine and thus SAM increases, SAM induces a temporal arrest in the translation process of AtCGS mRNA. This arrest occurs when the ribosome translates a domain called MTO1 (for methionine over-accumulation), a small conserved domain in the N-terminal region of AtCGS. As a result of the translation arrest, mRNA degradation occurs upstream of the stalled ribosome, resulting in the production of 5'-truncated RNA species leading to transcript decay [30].

The role of the MTO1 domain in the regulation of CGS level during the development and growth of plants is not fully clear. There is no correlation between high methionine levels and low AtCGS mRNA levels in transgenic *Arabidopsis* plants overexpressing the endogenous AtCGS [23]. Moreover, overexpression of AtCGS in tobacco [32, 33], potato [34-36] and alfalfa plants [37] showed a positive correlation between AtCGS level and methionine content. Whether these observations are due to variations in MTO1 machinery between *Arabidopsis* and other plant species, or are the result of the overexpression itself needs to be elucidated. Questions remain regarding the nature of MTO1 regulation in other species. While exogenous application of methionine significantly reduces the expression level of CGS in Arabidopsis, tomato [38] and transgenic tobacco expressing AtCGS [33], the potato CGS1, which shares 97% identity with tomato CGS, was not affected by application of exogenous methionine [39]. Moreover, overexpression of potato CGS in potato plants, even though it increases enzyme activity significantly, does not cause an elevation in methionine level, implying that CGS in potato plants does not play a major role in controlling methionine level [40]. This observation suggests that potato plants miss some elements required for post-transcriptional regulation of CGS [39]. Apparently, in potato, other elements are involved in the regulation of Met levels.

Alignment analysis of bacteria, yeast and plants had shown that plants possess an N-terminal region comprised of ~100 amino acids (not including the plastid transit peptide), which does not exist in bacterial enzymes. This N-terminal region is not essential for the catalytic activity of CGS but it influences the levels of methionine and its associated metabolites, such as ethylene and dimethylsulfide [32]. Another study performed on Arabidopsis revealed an additional regulatory domain within the N-terminal region of the AtCGS. The additional domain is adjacent to the MTO1 domain and contains 90 nt [33]. Arabidopsis has only one gene for CGS (AtCGS), but it has two tanscripts and two proteins. One is the full-length AtCGS and the second is identify to the full-length but has deletion of the 90 nt in the N-terminal region. Thus it was suggested that the AtCGS form that lacks 90 nt (named "deleted" form of AtCGS) originates from the full-length AtCGS. Transgenic tobacco plants that over express the deleted form of AtCGS show that this form is not subject to feedback regulation by high methionine content [33]. As a result, these transgenic plants accumulate five-fold higher levels of methionine compared to plants expressing the full-length AtCGS, and about nine-fold compared to wild-type plants [33, 41]. A high level of methionine (up to six-fold compared to wild-type plants) was also found in tuber potatoes over expressing this deleted form of AtCGS [42]. This high elevation encouraged us to expresses this form of AtCGS in the forge legume, alfalfa, which has low methionine content. In the leaves of these plants, the level of soluble methionine increased up to 32-fold and the methionine incorporated into the water-soluble protein fraction increased by 2.2-fold, compared to wild-type plants. As a result, the methionine percentage increased from 1.63% to 3.65% in transgenic lines. Taken together, these results suggests the MTO1 region plays a major role in controlling the CGS expression level, but additional *cis* or *trans* elements are also involved in the regulation.

The roles of the MTO1 and the deleted domains in the regulation of AtCGS are still being studied. We assume that in addition to the MTO1 and the deleted regions, additional domains with distinct functions are contained within AtCGS N-terminal region. This assumption is derived from the observation that transgenic tobacco plants overexpressing AtCGS lacking the N-terminal region have severely abnormal phenotype. This includes stunted growth, slow developmental rate, loss of apical dominance, and narrow, greener, and curly leaves [32]. In some of these plants, the apical meristems and leaf primordium turn brown and dry, and flower buds are produced in only few of the transgenic plants, that fall soon after the abscission zone is formed, causing total sterility. The transgenic plants produced many secondary stems and were able to survive for more than 2 years. In addition, the transgenic plants emitted high levels of ethylene and dimethylsulfide [32]. However, plants overexpressing the full length AtCGS or its deleted form show only slight morphological phenotype, although they have similar or 9-fold higher soluble methionine levels, respectively, compared to plants expressing the AtCGS without its N-terminal region. Speculation on the function of this region also derive from the observation that under conditions of starvation for folates for prolonged periods, this region is removed by proteolytic cleavage, leaving the enzyme active in a similar manner to the full length CGS [43]. Any connection between low folate and CGS requires further studies.

The main control on AtCGS is by the regulation on its transcript level through the MTO1 and the deleted domains. However, additional factors regulate the level of AtCGS. Threonine in *Arabidopsis* [37, 44] and the SAM-derived hormone, ethylene in tomato [38] enhance the transcript level of CGS. CGS expression level in potato plants is up regulated in light [45], and we have recently found that the expression of AtCGS is also up regulated by light through sucrose (R. Amir, unpublished results). Tight regulation of CGS is vital for plant survival, as was demonstrated by the severe morphological phenotypes that usually appear in plants having high or low levels of CGS. Down-regulation of CGS in *Arabidopsis* led to severe growth retardation; these plants had reduced apical dominance and their floral organs were stacked. In addition, chlorophyll content was reduced significantly [23, 46, 47]. Abnormal phenotype was also observed in *Arabidopsis* and potato overexpressing the CGS and thus having higher methionine contents [23, 42]. However, this severe abnormal phenotype is not always manifested, as was reported for alfalfa [37, 44], potato [34-36, 40, 48] and tobacco [41]. Low germination rate was observed in tobacco seeds expressing AtCGS in a seed-specific manner. These seeds have high methionine content and normal morphological phenotype (R. Amir unpublished results). These results suggest that high methionine content, or one of its associated metabolites, affect the normal development and growth of the plants.

The Regulatory Role of Threonine Synthase

Several lines of evidence have shown that methionine biosynthesis is also regulated by threonine synthase (TS), the last enzyme of the threonine biosynthesis pathway in plants. This enzyme competes with CGS for their common substrate, *O*-phosphohomoserine (OPH) (Fig. 1) [39, 49]. Analyses of transgenic and mutant plants have shown that this competition regulates the level of methionine, since reduction in TS levels using an antisense approach in potato and *Arabidopsis* plants caused a significant increase in methionine level by up to 239-fold and 47-fold, respectively [37, 44, 50]. In addition, the TS *MTO2* mutant shows significantly higher methionine levels (22-fold) [51].

Several observations suggest that SAM, the first catabolic product of methionine, is one of the factors controlling the flux towards threonine and methionine biosynthesis pathways: SAM negatively regulates the expression level of CGS [30] and *in vitro* studies show that SAM positively regulates TS activity by an allosteric mechanism that increases the affinity of TS to OPH [52, 53]. This implies that methionine regulates its own synthesis through SAM.

When high levels of methionine and consequently SAM, are produced, CGS level is down regulated and TS activity is enhanced, causing reduced OPH availability for methionine synthesis [52, 53]. However, other evidences suggest that this scenario does not exist under *in vivo* conditions. Analyses of some mutants have demonstrated that TS activity and threonine levels are not affected by SAM content. For example, it was found that in the *mto1* mutant in which the CGS gene was mutated, free methionine levels increased 40-fold and SAM levels increased three-fold, whereas free threonine level was not altered compared to wild-type plants [26, 51]. In the *mto3* mutant having a mutation in the SAM synthase 3 gene, methionine increased more than 200-fold and SAM decreased by 35%, but free threonine was again not affected [54]. Moreover, in the *mmt* mutant, an opposite relationship prevailed between SAM and threonine content, since this mutant has a higher level of SAM and a lower level of threonine [13]. These results are intriguing, as they imply that SAM may not lead to the activation of TS *in vivo*.

Regulation of Carbon/Amino Flow into Methionine Synthesis

The studies described above suggest that the carbon-amino skeleton derived from aspartate through OPH limits the levels of both methionine and threonine. In order to study this further, homoserine kinase, the enzyme that produces OPH (Fig. 1), was overexpressed in *Arabidopsis* and potato plants [37, 44, 55, 56]. Expression of this enzyme does not lead to higher levels of methionine and threonine in these transgenic lines [37, 44, 55,56]. However, when wild-type and transgenic potato and *Arabidopsis* plants overexpressing homoserine kinase were fed with homoserine, one metabolite upstream to OPH (Fig. 2), the levels of both amino acids increased significantly [37, 44, 55, 56]. Moreover, a marked and significant increase in methionine content (a 180-fold) occurred when *Arabidopsis* plants overexpressing the AtCGS were fed with homoserine [55]. These results suggest that under physiological conditions, the AtCGS and TS are substrate limited, and the levels of these amino acids is tightly regulated by the flux of the carbon-amino skeleton towards their synthesis pathways.

Pervious results also suggest that the carbon-amino skeleton limits threonine synthesis. Overexpression of bacterial feedback-insensitive aspartate kinase (AK) [57, 58] in tobacco, *Arabidopsis* and alfalfa plants led to significantly higher threonine level while the methionine content, whose biosynthesis pathway diverges from this branch, was not significantly changed [57-60].

To further study if the carbon-amino skeleton limits the synthesis of methionine, we have produced transgenic tobacco plants overexpressing both, AtCGS [32] and bacterial feedback-insensitive AK. However, the level of methionine in these plants does not increase significantly compared to plants over expressing the AtCGS alone [41]. This result can be explained by the feedback-inhibition regulation mediated by SAM on the transcript level of AtCGS that occurs when methionine level increases beyond a certain threshold [27, 28]. To test this further, we have produced plants overexpressing the bacterial AK and the deleted form of AtCGS, which is methionine feedback insensitive. In these plants, methionine level has significantly increased compared to plants expressing only this form of AtCGS (about 4.5-fold higher), in comparison to wild-type plants, the increase in methionine contents was about 110- to190-fold. Threonine levels doubled in these plants compared to wild-type plants [41]. These results suggest that the carbon-amino skeleton limits methionine synthesis, but this can hardly be detected under normal physiological conditions since the regulatory role of SAM on the expression level of AtCGS has a stronger effect.

The Regulatory Role of Cysteine, the Sulfur Donor, in the Methionine Biosynthesis Pathway

The results obtained above suggest that the carbon-amino skeleton substrate of CGS limits the biosynthesis of methionine and thus its level. However, CGS also uses cysteine as a substrate. In order to determine if cysteine content limits methionine synthesis, methionine content was analyzed in transgenic tobacco plants having higher levels of cysteine due to overexpression of the yeast *Met25* gene coding the last enzyme of the cysteine biosynthesis pathway, *O*-acetylserine (thiol) lyase (yOASTL). Methionine level did not change significantly in these plants [61]. However, overexpressing this yeast gene in transgenic flax plants resulted in a 3.8 fold increase in the level of methionine compared to non-transgenic flax [62]. In both transgenic plants, tobacco and flax, the level of the cysteine metabolite, glutathione, increased significantly. Higher levels of glutathione were also observed in other transgenic plants having high levels of cysteine (e.g., [63-67]. This suggests that cysteine is channeled towards glutathione synthesis, whereas the ability of high cysteine content to affect methionine content depends on the plant species.

The observation that transgenic plants overexpressing the AtCGS without its N-terminal region, and thus having higher levels of methionine emit significantly higher levels of sulfur-containing metabolites such as methanethiol, dimethylsulfide and carbon disulfide [32, 69], suggest again that sulfur content as well as cysteine do not limit methionine synthesis in plants.

However, the relationship between sulfate availability and methionine content in plants is not yet clear. It was reported that over a broad time range of sulfate deprivation, methionine levels do not change significantly [69, 70]. The plants were able to keep methionine level constant, which is always low in plants. These results suggest that factors other than sulfur content are involved in methionine homoeostasis in plants during sulfur starvation. One factor that might assist in maintaining the methionine homoeostasis in plants could be the methionine-recycling pathways that recycle methionine-moieties to regenerate methionine (described below). Indeed, it was found that the transcript levels of SAM synthase and adenosylhomocysteine hydrolase (SAH), two enzymes involved in recycling methionine moieties (Fig. 2), are induced during sulfur starvation [70]. Another option, which needs to be studies, is that during sulfur starvation, proteins are proteolyzed, keeping soluble methionine content stable.

The Regulatory Role of Cystathionine β-Lyase

Cystathionine β-lyase, the second enzyme in the methionine biosynthetic pathway, converts cystathionine to homocysteine. Cystathionine β-lyase activity is essential for plant growth and development as demonstrated by the compromised phenotypes of a tobacco mutant [71] and of transgenic potato plants expressing antisense of cystathionine β-lyase [72]. In addition to abnormal morphological phenotype characterized by bushy growth, small leaves and small tubers, the plants had increased levels of cysteine, cystathionine and homoserine, which are intermediate metabolites upstream to cystathionine β-lyase. The level of free methionine was reduced, but unexpectedly the level of homocysteine, the product of cystathionine β-lyase, was increased in the transgenic plants. These findings could possibly be explained by the induction of an additional pathway capable of producing homocysteine in plants, which is the direct sulfhydrylation. This pathway performed by CGS that can produce homocysteine from OPH using sulfide instead of cysteine as the sulfur donor. It was suggested that when the regular methionine biosynthetic pathway is blocked by a decrease in the cystathionine β-lyase activity in transgenic plants, it may enhance the activity of CGS in this direct pathway. Giovanelli et al.. (1985), estimate that this pathway contributes only 3% of total homocysteine synthesis in plants under normal growth conditions. However, *in vitro* studies using complementation tests in *E. coli* mutants, using the AtCGS indicated that this reaction can account for a much higher proportion of homocysteine synthesis, operating at up to 60% of the activity of CGS [73].

The cystathionine β-lyase expression level is apparently subject to regulatory control by methionine or one of its associated metabolites. In Arabidopsis plants with reduced level of CGS, the expression level and enzymatic activity of cystathionine β-lyase had increased by two-fold [46, 47]. Overexpression of cystathionine β-lyase in transgenic potatoes did not increase metabolite flow towards methionine, and did not affect the other metabolites tested [74]. Thus, cystathionine β-lyase seems to have only minor effect on the flux towards methionine.

Recently it was found that cystathionine β-lyase level is up-regulated during drought stress in tomato. The researchers suggest that this may affect SAM metabolism, and thus affect the SAM metabolites such ethylene, polyamines, nicotianamine and the rate of some methyl transfer enzymes [75].

The Regulatory Role of Methyl Flow in Methionine Synthesis

Methionine synthase, the last enzyme of the methionine biosynthesis pathway, catalyzes the methylation of homocysteine to methionine by using 5-methyltetrahydrofolate as a methyl group donor. This enzyme is also required for the regeneration of methionine after the donation of a methyl group by SAM, generating *S*-adenosylhomocysteine that is converted to homocysteine, and thus is recycled to form methionine (Fig. 2). Three forms of methionine synthase were found in *Arabidopsis*, two in the cytosol and one in the chloroplasts, suggesting that methionine is synthesized both in the cytosol and in chloroplasts. The two cytosolic methionine synthase enzymes suggest that homocysteine produced in the chloroplasts is transferred from this compartment to the cytosol, where it is used by methionine synthase to form methionine [76]. The chloroplast enzyme synthesizes methionine in the chloroplasts, to support chloroplastic protein synthesis and methionine metabolism [77]. Further studies are required in order to reveal whether the methionine synthesized in the chloroplast can be exported to the cytosol, thus contributing to methionine metabolism in this compartment.

As expected from its key role in methionine recycling pathways, methionine synthase does not play any regulatory or limiting role in methionine biosynthesis, as suggested by over expressing the gene encoding this enzyme in potato plants. Although the level of this enzyme has increased significantly in the transgenic plants, methionine content has not changed significantly [39, 78].

The transcrip expression level of methionine synthase is regulated by light, sucrose or other photoassimilates, reaching its highest level at midday, while maintaining a basal level at night. However, the fluctuations in transcript level do not correlate with changes in methionine synthase protein content, which remained constant over the whole day [76]. Methionine synthase content is strongly elevated during seed germination in Arabidopsis, most probably to support the production of the plant hormone, ethylene during germination [79]. Similarly, methionine synthase level was highly elevated during ethylene-dependent fruit ripening in tomatoes, indicating that it may be regulated by ethylene [80]. However, understanding the co-regulation of methionine synthase with cystathionine β-lyase and CGS in various tissues such as seeds and fruits, demands further studies.

The Regulatory Roles of Lysine and Threonine in Determining Methionine Levels in Plants

Two major branches exist in the aspartate family of amino acids, the lysine branch and the threonine, methionine and isoleucine branch (Fig. 1). The activity of AK, the first enzyme of aspartate family, controls the flux towards the amino acids in the family. However, the regulation of the carbon-amino skeleton flux within these two branches is quite complicated, and metabolites from one branch affect flux towards the other branch [81]. Application of both threonine and lysine leads to methionine starvation since they both inhibit AK activity, the first enzyme of the aspartate family [37, 44, 55] (Fig. 1). However, threonine and lysine, when applied separately, lead to methionine accumulation. Using feeding experiments and transgenic plants, it was found that high levels of threonine enhance the expression levels of AtCGS and thus lead to an increase in methionine content [37, 41, 44]. However, studies using the *in vitro* transcription/translation system suggest that threonine does not directly affect the level of AtCGS [41], thus implying that other metabolites connected to high threonine level affect AtCGS expression level.

In addition to threonine, previous studies have also shown a positive correlation between methionine and lysine content in plants. Transgenic tobacco plants overexpressing AtCGS and feedback-insensitive dihydrodipicolinate synthase (bDHDPS), the key enzyme of lysine biosynthesis pathway [82, 83], show that the content of methionine and SMM is significantly elevated compared to those expressing AtCGS alone. In these plants, the level of lysine remained the same as in plants over expressing bDHDPS alone [84]. In plants overexpressing both heterologous genes, the expression level of AtCGS increases significantly, while the level of SAM is reduced. *In vitro* transcription/translation studies have shown that high levels of lysine down-regulate the transcript expression level of SAM synthase, the enzyme that converts methionine to SAM. This leads to a reduction in SAM, which negatively regulates the level of AtCGS transcript [28]. Therefore, the result of elevated lysine levels is an increase in the expression level of AtCGS and consequently an elevation in the level of methionine (Fig. 1, dotted line) [84]. Methionine level can also be enhanced by a reduction in the flux towards SAM and its metabolites [22]. Similar relations between methionine and lysine, were also described in transgenic barley plants that constitutively express bDHDPS. In these plants the level of lysine had increased 14-fold and the level of methionine increased 8-fold [85]. Similarly, seeds of three sets of transgenic *Arabidopsis* plants exhibiting significantly higher levels of lysine (80-fold) also show high methionine content (51-fold) compared to wild-type seeds [86, 87].

When considering wild-type plants, the question remains as to the conditions under which lysine accumulates to high levels and thus modulates methionine content and its metabolites. The levels of soluble lysine were found to increase in young leaves and at shoot apices where it is required for protein synthesis. As methionine is generally converted rapidly to SAM and its other metabolites [14, 22], such an effect of lysine on the level of SAMS enzymes may reduce the conversion rate of methionine to SAM. This increases the level of soluble methionine, which is required together with lysine for protein synthesis in the young tissues.

The results described above imply that the two-methionine "brothers" in the aspartate family, threonine and lysine, whose biosynthesis pathways compete with the methionine pathway for the carbon/amino skeleton, regulate methionine level, keeping methionine homeostasis. This regulation *in planta* ensures that if a mutation occurs or conditions leading to higher levels of threonine or lysine exist, the level of methionine will not be significantly altered. Regulating methionine level is important due to the major role that methionine plays in protein synthesis and plant metabolism.

2. Methionine Recycling Pathways in Plants

Additional factors that might assist in maintaining methionine homoeostasis in plants are the methionine-recycling pathways. Three recycling pathways can regenerate the moieties comprising methionine following the formation of SAM metabolites:

The methyl cycle: The utilization of the methyl group of SAM in transmethylation reactions leaves *S*-adenosylhomocysteine. *S*-adenosylhomocysteine, is a strong competitive inhibitor of the SAM-dependent methylases, therefore it must be removed for proper functioning of these enzymes [24]. This reaction is catalyzed by *S*-adenosylhomocysteine hydrolase (SAHH) that convert *S*-adenosylhomocysteine to homocysteine (the second intermediate metabolite of methionine biosynthesis pathway). The combination of homocysteine and methyl group by the cytosolic form of methionine synthase regenerates methionine [8] (Fig. 2).

The Yang cycle: in this cycle, the methyl-thio group of methionine is recycled and the Yang cycle metabolites are used as precursors for the synthesis of essential metabolites in plants: ethylene, polyamines, and nicotianamine. Disruption of the Yang cycle in the Arabidopsis *mtk* mutant resulted in reduced SAM level compared to wild type at sulfur-limiting conditions. When the Yang cycle knockout and ethylene overproduction were combined in the *mtk/eto3* double mutant, a reduced capacity for ethylene synthesis was observed in the seedlings. The researchers conclude that the Yang cycle contributes to the maintenance of SAM homeostasis, especially when *de novo* SAM synthesis is limited. In addition, the data also showed that this cycle is required to maintain high rates of ethylene synthesis [94]. However, additional evidence suggests that in addition to recycling the methionine moieties via the Yang cycle, the *de novo* synthesis of methionine is required when high rates of ethylene production are induced [38]. This suggestion is based on the observation that the transcript expression level of CGS and the level of methionine positively correlate to ethylene production during the ethylene burst of the climacteric ripening of tomato fruit and during leaf wounding upon ethylene emission [38].

The SMM cycle: SMM cycle is unique to plants. In this cycle SMM is formed from methionine via the activity of SAM:Met S -methyltransferase (MMT). SMM can be reconverted to methionine by donating a methyl group to homocysteine in a reaction catalyzed by homocysteine *S*-methyltransferase (HMT) [4] (Fig. 2). Each turn of this SMM cycle consumes and then regenerates two methionine molecules while hydrolyzing ATP, and thus this cycle is considered futile. Several possibilities have been attributed for the function of SMM in plants: (a) SMM is a transport form of sulfur that mediates the long-distance transport of labile methyl moieties and reduces sulfur [22]; (b) SMM functions to convert methionine moieties found in SAM back to free methionine. At this way the SMM cycle keeps the free methionine pool from being depleted through SAM synthesis [88]; (c) SMM controls the SAM level in plant tissue. This control was hypothesized since plants lack the negative feedback loops that regulate SAM levels in other eukaryotes [14].

In silico model developed from radiotracer kinetic measurements indicated that the SMM cycle helps to stop the accumulation of SAM, rather than to prevent depletion of free methionine [14]. Eliminating the SMM cycle in Arabidopsis by knockout mutants of MMT demonstrated that SMM is unessential for growth or development in normal conditions. This *mmt* mutant had increased SAM level and methylation ratio while free methionine and thiol pools were not significantly changed. These finding supported the assumption that the SMM cycle contributes to the regulation of SAM levels [13]. Further support came from studies showing that Arabidopsis cells that were fed with methionine and therefor had elevated amounts of methionine were able to maintain a low and steady concentration of SAM while the level of SMM increased [89].

While the role of SMM in plants metabolism has not fully clarified yet, there is evidence that SMM plays a major role in methionine synthesis in seeds. Based on these studies, a model was developed, suggesting that methionine that is synthesized in vegetative tissues, mainly in leaves, is converted to SMM by MMT enzyme. The SMM is then transported to seeds where it is converted by HMT to methionine. The evidence supporting this model is: (i) The high level of methionine found in rosette leaves of Arabidopsis mutants *mto1-1* and *mto2-1* is reduced as the mutants begin to flower. Relatively high methionine content is detected in the reproductive organs of these mutants [26]. These findings suggest that at the onset of reproductive growth, methionine is translocated to sink organs and most probably not synthesized in situ in these organs. (ii) By using radioactive methionine, it was found that methionine can be exported from leaves in the form of SMM, which is the major sulfur metabolite in the phloem of different plants including Arabidopsis [90]. Measurements and calculations have farther suggested that in leaves of wheat, the flux is mainly from methionine to SMM, and in the seeds from SMM to methionine [90]. (iii) Analysis of Arabidopsis *hmt2* mutant, revealed that this mutant accumulates high level of methionine in seeds and high levels of SMM in the upper flower stalk (stems, cauline leaves, flowers, silique hulls and seeds) [91]. Using [^{13}C]SMM and [^{13}C]Met it was shown that increased accumulation of SMM at the top of the *hmt2* flower stalk increased the transport of SMM to seeds. In seeds, a portion of this transported SMM is converted to methionine by seed HMT isozymes (other than HMT2) [91]. (iv) Proteome and transcriptome analyses of developing

Medicago truncatula suggest that methionine is synthesized in the seed coat from the pool of SMM and is subsequently transported to the embryo where it is assembled into proteins and accumulated [92]. This occurs at a late stage of seed development due to increased demand for protein synthesis during seed filling. Together, these results support a model whereby SMM serves as an intermediate during transport of methionine from vegetative tissue to seeds in Arabidopsis.

However, further studies are required to reveal the importance of SMM cycle since it was shown that the lack of SMM cycle in *mmt* mutant in Arabidopsis does not alter the levels of methionine, thiols and seed sulfur compared to wild-type seeds [13]. In addition, transgenic tobacco plants overexpressing AtCGS and the bacterial feedback-insensitive form of aspartate kinase, which exhibit about 170-fold higher methionine and about 30-fold higher SMM content in their leaves, have only slightly higher methionine content in their seeds [41]. These results suggest that SMM is not the main source for methionine synthesis in seeds. Indeed other studies suggest that methionine is synthesized de novo by CGS, since seed specific expression of bacterial AK led to significant higher levels of methionine [35, 93].

3. Methionine Catabolism in Plants

The Role of SAM Synthase in Controlling Methionine Content

The conversion of methionine to SAM is catalyzed by SAM synthase (SAMS) in an ATP-dependent reaction. Studies of the metabolic fates of methionine using the aquatic plant *Lemna paucicostata* indicated that about 80% of methionine is converted to SAM and its associated metabolites (i.e. ethylene, polyamines, SMM, methyl group donor etc.) and 20% is used for protein synthesis [95, 96] (Fig.1). However, in mature Arabidopsis rosette leaves, about half of the soluble methionine is converted to SAM and SMM, and half is used for protein synthesis [14].

SAMS plays a major role in controlling methionine levels in plants. Reduction in the expression level of SAMS leads to significantly higher levels of methionine in plants, as shown by several mutants and transgenic plants. For example, the *Arabidopsis* mutant (*mto3*) exhibiting reduced SAMS3 activity had over 200-fold methionine content compared to wild-type plants [54, 97]. This high level of methionine was associated with a low level of SAM, and as a result, the lignin content, one of the major metabolic sinks for SAM, decreased by 22% compared to wild-type plants [54]. A more severe phenotype appeared in *Arabidopsis* plants with silenced SAMS expression. In these plants, the level of methionine has increased 250-fold compared to wild-type plants, but severe abnormalities were observed [23]. A severe abnormal phenotype also appeared in tobacco plants where the SAMS level was reduced significantly [68]. This suggests that a reduction in the expression level of SAMS could significantly increase the level of methionine, but due to the importance of SAM in plant metabolism, such a reduction would lead to severe abnormal phenotypes. Alternatively, high levels of methionine can be toxic to the plants. Unlike the former results, over expression of SAMS in *Arabidopsis* did not lead to altered levels of methionine and SMM [23].

The Role of Methionine γ-Lyase

Several lines of evidence suggest that methionine can be degraded in plants. These include findings that plants having high methionine content emit volatile compounds containing sulfur and methyl moieties, such as methanthiol, dimethyl disulfide, or dimethyl sulphide; this suggests that these metabolites are methionine degradation products [32, 68]. The enzymes and pathways responsible for the production of these volatiles and their physiological significance are still unknown.

One enzyme involved in methionine catabolism is methionine γ-lyase (Fig. 1). This enzyme is located at the cytoplasm and produces 2-ketobutyrate, methanethiol, and ammonia from methionine [89, 98]. The researchers found that this cytosolic enzyme is abundant in all plant organs, except in the seeds. Western blot studies indicated that methionine γ-lyase is expressed under standard growth conditions and is strongly induced when the cells accumulate methionine [89]. The enzyme has a relatively high K_m value for methionine (~10 mM), indicating that this pathway operates preferentially when methionine has accumulated above a certain value in the cytoplasm. Under sulfate limitation, knockout of the single *Arabidopsis* methionine γ-lyase gene significantly increased methionine content (9-fold) and SMM content in leaf, but there was no effect on methionine levels under normal growth conditions [98]. This finding suggests that this catabolic pathway plays a role during sulfate starvation. Nevertheless, it was shown that under this stress the level of methionine is not altered in wild type plants [70], therefore further studies are required to reveal the role of this enzyme.

Questions remain regarding the role of methanthiol, the catabolic product of methionine, in plant metabolism. Rebeille et al.. (2006) suggest that 75% of the methanthiol is released from cells as a volatile compound and 25% react with *O*-acetylserine (the metabolite of the cysteine biosynthesis pathway) to produce *S*-methylcysteine. In Arabidopsis plants fed with radioactive methionine on a low sulfate medium, about 9% of the radioactive methionine appeared as cysteine in proteins. The incorporation into cysteine was significantly (30%) less in the methionine γ–lyase knockout mutant [98]. The researchers therefore assume that the methanthiol pathway is alternative to the reverse transsulfuration pathway operating in certain bacteria [99] in which methionine is converted to cysteine.

α-Ketobutyrate, the second catabolic product of methionine γ-lyase, can be transported from the cytosol to plastids and integrate to the isoleucine biosynthesis pathway [100]. A relationship between high methionine and isoleucine contents was also reported in potato, tobacco and *Arabidopsis* plants [41, 42, 50, 89]. The biosynthesis of isolecucine through the activity of methionine γ-lyase, plays a significant role under abiotic stresses, such as osmotic, drought and salt stresses [100. 101]. The reasons for this are not yet clear.

The Role of Enzymes Leading to Glucosinolates Synthesis

Methionine is also catabolized by methyl thioalkylmalate synthases (MAM) to produce glucosinolates (GSLs), a class of plant secondary metabolites found in Arabidopsis and other genera of the Brassicaceae [102, 103]. GSLs exhibit repellent activity against herbivore insects and pathogens. These compounds are sulfur-rich plant metabolites that are produced from methionine or from aromatic amino acids, and can represent up to 30% of the total sulfur content of plant organs [10]. Methionine-derived GSLs with diverse side chains of various lengths are the major GSLs in *Arabidopsis*. It was recently reported that in the knockout lines of the *atleuc1-1* gene, one of the enzymes of the methionine-derived GSLs synthesis pathway, the levels of six methionine-related metabolites, SAM, methionine, SMM, methionine sulfoxide, *O*-acetylserine and 5′-deoxy-5′-methylthioadenosine (a byproduct in the biosynthesis of spermine and spermidine), are significantly elevated [104]. Similarly, *bcat4* knockout mutant shows that the levels of aliphatic GSLs are reduced by approximately 50% while the levels of free methionine and SMM are significantly increased [103]. This result suggests that when the metabolic flow from methionine to methionine-GSLs biosynthesis is blocked, redirection to the primary methionine metabolism occurs.

4. The Importance of Methionine in Human and Livestock Nutrition

Aside from the importance of methionine for plant metabolism, methionine is also a fundamental metabolite in animal cells. Methionine belongs to the "essential amino acids" that humans and livestock cannot synthesize and thus must obtain them from their diets. Animals can convert methionine to cysteine, but not conversely, thus defining methionine as an essential amino acid. Cysteine, however, considered a "conditionally" essential amino acid. Therefore, methionine can supply the complete requirement for sulfur amino acids in the diet [19]. Plants producing both amino acids are the ultimate source of sulfur amino acids in the food of animals and humans. Cysteine and methionine are required for protein synthesis, in which they are nearly always directly involved in the catalytic, structural, or electrochemical functions of proteins, and give rise to key enzymatic cofactors [105]. In addition to their role in protein synthesis, the sulfur amino acids are the primary route for the body to incorporate sulfur into a vast array of molecules. The metabolism of these amino acids is also linked to pathways involved in glutathione (an important antioxidant and xenobiotic detoxification molecule), SAM, the universal methyl donor, and folic acid. They also serve as metabolic precursors in the production of glycosamineglycans, an important component in connective tissues like cartilage and skin [106]. Likewise, the synthesis of vitamins containing sulfur, such as thiamine, biotin and Co-enzyme A depend on the availability of these amino acids.

The nutritional value of plants, which are the world's most dominant sources of human food and animal feed, is predominantly inadequate due to their limited content of some essential amino acids. While cereal grains are limited in the levels of tryptophan, lysine and threonine; legumes (e.g., soybean, bean, chickpea, pea, alfalfa, and lentil) are mainly limited in the contents of sulfur amino acids, methionine and cysteine [19, 37, 107]. The biological value of a plant-based diet with limited methionine content can be equivalent to only 50% to 75% of that of a diet with balanced essential amino acids. In developing countries in which plant-derived foods are predominant, this can lead to non-specific signs of protein deficiencies in humans, such as lowered resistance to disease, decreased blood proteins and retarded mental and physical development in young children [108]. In addition, because methionine is one of the four main dietary sources of methyl groups, its deficiency can be associated with methylation-related disorders, such as fatty liver, atherosclerosis, neurological disorders and tumorigenesis [109-111]. Methionine, through SAM, also influences DNA synthesis and repairs the expression of genes. Methionine deficiency has been associated with DNA fragmentation and strand breaks [112].

Deficiency of essential amino acids is rare in humans in developed countries due to the significant consumption of livestock products, which contain adequate levels of these essential amino acids. However, deficiencies in these essential amino acids in plant-derived feeds for farm animals do represent a major economic disadvantage, limiting both livestock growth and their productivity, such as reduced milk production and its protein concentration, decreased meat quality and reduction in egg size and numbers (e.g., [113-116]. To meet the requirements of monogastric animal diets, methionine is added in a synthetic form to animal-based diets in many Western countries. It is also added to processed soybean products (e.g., soybean 'milk' and tofu) for human consumption.

Due to the importance of methionine in human food and animal feed, many efforts have been made to produce plants having higher methionine content. However, plant breeding methods and selection of mutants have yielded only limited success in increasing the level of sulfur amino acids [117]. Recent developments in recombinant DNA technology, plant tissue culture and in vitro regeneration are proposing new ways of increasing the level of essential amino acids, including methionine, by manipulating existing genes and/or introducing foreign genes into plants.

Increasing Methionine Content in Plants

Two main strategies have been used to increase methionine content via gene transfer technology: manipulation of the methionine biosynthetic pathway and the creation of additional protein "sinks" for methionine storage (reviewed by [118]. Both molecular approaches have provided new insights into the control of methionine level in plants, and in some cases have yielded significantly higher levels of methionine and consequently improved plant nutritional value.

The expression of genes encoding methionine-rich seed storage proteins fused to a constitutive promoter or seed-specific promoter was tested in different crops and model plants. In some cases, these heterologous proteins had accumulated, leading to higher methionine content [19, 118]. However, in most of these cases the elevation of heterologous protein was at the expense of the other sulfur-rich endogenous proteins, as well as other sulfur compounds in the cells (reviewed by [118]. This indicates that methionine availability limits the synthesis of methionine rich-proteins. It was therefore suggested that in addition to expressing methionine-rich storage proteins, it is desirable to increase the soluble methionine content by manipulating the methionine biosynthesis pathway or reducing its catabolism [17, 119].

Several approaches have been tested, including overexpression of CGS, reducing the expression levels of TS and SAMS, and overexpressing combinations of AK/CGS, and of CGS/bDHPS, as described above (also reviewed by [17, 18, 20, 107]. The most promising approaches are the enhancement of the level of methionine by manipulating its biosynthesis pathway. Expression of the deleted form of AtCGS alone or together with AK, led to significantly higher methionine levels and methionine together with threonine, respectively

[41]. Expression of AtCGS with bDHPS led to higher level of both methionine and lysine [33]. However, the reduction of TS levels by the expression of antisense constructs, although it led to higher level of methionine,significantly reduced the level of the essential amino acid threonine, and led to severely abnormal phenotype [41]. Similarly low level of SAMS led to severely deformed phenotype [23, 90]. Thus, these two strategies are not suitable for the enhancement of the level of methionine from a biotechnological point of view.

To further define if indeed the level of soluble methionine limits the production of sulfur-rich proteins in vegetative tissues, transgenic alfalfa plants constitutively expressing the β-zein were crossed with those expressing the deleted form of AtCGS [37]. Compared to plants expressing only the β-zein, those co-expressing both transgenes showed significantly enhanced levels of β-zein concurrently with a reduction in the level of soluble methionine when compared to plants expressing the AtCGS alone. This implies that more soluble methionine incorporated into the β-zein in the crossed plants [120, 121].

Transformation of potatos with β-zein and AtCGS has caused the level of methionine to increase six-fold. However, the transgenic plants exhibited severe growth retardation, changes in leaf architecture and a 40-60% reduction in tuber yield. Furthermore, the color of the transgenic tubers has altered due to the reduced amounts of anthocyanin pigments [42]. Therefore, additional studies are required in order to obtain crop plants having higher methionine content with normal phenotype. Morphological changes are less expected in seeds where methionine metabolism is supposed to differ from vegetative tissues. However, knowledge about methionine synthesis and its metabolism in seeds is still limited and requires further study [17, 20, 91]. We believe that the knowledge accumulated regarding vegetative tissues described in this review will help in the manipulation of methionine metabolism in seeds, and thus help in the production of seed-crop plants having higher levels of methionine and consequently better nutritional quality.

Acknowledgments

The study on methionine metabolism was supported by the Israel Science Foundation (grants 566/02-1 and 231/09).

References

[1] Matilla, A. J., Ethyelene in seed formation and germination. *Seed sci res.* 2000, 6, 111-126.

[2] Yang, S. F.; Hoffman, N. E., Ethylene biosynthesis and its regulation in higher plants. *Annu Rev Plant Physiol.* 1984, 35, 155-189.

[3] Pang, X. M.; Zhang, Z. Y.; Wen, X. P.; Ban, Y.; Moriguchi, T., Polyamines, all-purpose players in response to environment stresses in plants. *Plant Stress* 2007, 1, 173–188.

[4] Droux, M., Sulfur assimilation and the role of sulfur in plant metabolism: a survey. *Photosynth Res* 2004, 79, (3), 331-48.

[5] Nikiforova VJ, G. B., Kempa S, Adamik M, Willmitzer L, Hesse H, Hoefgen R., Towards dissecting nutrient metabolism in plants: a systems biology case study on sulphur metabolism. *J Exp Bot.* 2004, 55, 1861-1870.

[6] Hesse H, K. O., Maimann S, Zeh M, Hoefgen R., Current understanding of the regulation of methionine biosynthesis in plants. *J Exp Bot.* 2004, 55, 1799-1808.

[7] Ravanel S, G. B., Job D, Douce R., The specific features of methionine biosynthesis and metabolism in plants. *Proc. Natl. Acad. Sci. USA* 1998, 95, 7805-7812.

[8] Roje, S., S-Adenosyl-L-methionine: beyond the universal methyl group donor. *Phytochemistry* 2006, 67, (15), 1686-98.

[9] Kagan, R. M.; Clarke, S., Widespread occurrence of three sequence motifs in diverse S-adenosylmethionine-dependent methyltransferases suggests a common structure for these enzymes. *Arch Biochem Biophys* 1994, 310, (2), 417-27.

[10] Hirai, M. Y.; Sugiyama, K.; Sawada, Y.; Tohge, T.; Obayashi, T.; Suzuki, A.; Araki, R.; Sakurai, N.; Suzuki, H.; Aoki, K.; Goda, H.; Nishizawa, O. I.; Shibata, D.; Saito, K., Omics-based identification of Arabidopsis Myb transcription factors regulating aliphatic glucosinolate biosynthesis. *Proc Natl Acad Sci U S A* 2007, 104, (15), 6478-83.

[11] Gigolashvili, T.; Berger, B.; Mock, H. P.; Muller, C.; Weisshaar, B.; Flugge, U. I., The transcription factor HIG1/MYB51 regulates indolic glucosinolate biosynthesis in Arabidopsis thaliana. *Plant J* 2007, 50, (5), 886-901.

[12] Mudd, H.; H., D. A., The *S*-methylmethionine cycle in *Lemna paucicostata. Plant physiol.* 1990, 93, 623-630.

[13] Kocsis, M. G.; Ranocha, P.; Gage, D. A.; Simon, E. S.; Rhodes, D.; Peel, G. J.; Mellema, S.; Saito, K.; Awazuhara, M.; Li, C.; Meeley, R. B.; Tarczynski, M. C.; Wagner, C.; Hanson, A. D., Insertional inactivation of the methionine s-methyltransferase gene eliminates the s-methylmethionine cycle and increases the methylation ratio. *Plant Physiol* 2003, 131, (4), 1808-15.

[14] Ranocha, P.; McNeil, S. D.; Ziemak, M. J.; Li, C.; Tarczynski, M. C.; Hanson, A. D., The S-methylmethionine cycle in angiosperms: ubiquity, antiquity and activity. *Plant J* 2001, 25, (5), 575-84.

[15] Hanson, A. D.; Rivoal, J.; Paquet, L.; Gage, D. A., Biosynthesis of 3-dimethylsulfoniopropionate in Wollastonia biflora (L.) DC. Evidence that S-methylmethionine is an intermediate. *Plant Physiol* 1994, 105, (1), 103-10.

[16] Simo, R.; Grimalt, J. O.; Albaiges, J., Sequential method for the field determination of nanomolar concentrations of dimethyl sulfoxide in natural waters. *Anal Chem* 1996, 68, (9), 1493-8.

[17] Amir, R., Towards improving methionine content in plants for enhanced nutritional quality. . *Func Plant Sci Biotechnol* 2008, 2, 36-46

[18] Amir, R., Current understanding of the factors regulating methionine content in vegetative tissues of higher plants. *Amino Acids* 2010, 39, (4), 917-31.

[19] Tabe, L. M.; Higgins, T. J., Engineering plant protein composition for improved nutrition. *Trends Plant Sci* 1998, 3, 282-286.

[20] Amir, R.; Han, T.; Ma, F., Bioengineering approaches to improve the nutritional values of seeds by increasing their methionine content *Mol Breed* 2012, 29, 915-924.

[21] Azevedo, R. A.; Arruda, P.; Turner, W. L.; Lea, P. J., The biosynthesis and metabolism of the aspartate derived amino acids in higher plants. *Phytochemistry* 1997, 46, (3), 395-419.

[22] Giovanelli, J. M., S.H. Datko, A.H., Quantitative analysis of pathways of methionine metabolism and their regulation in *Lemna. Plant Physiol.* 1985, 78, 555-560.

[23] Kim, J.; Lee, M.; Chalam, R.; Martin, M. N.; Leustek, T.; Boerjan, W., Constitutive overexpression of cystathionine gamma-synthase in Arabidopsis leads to accumulation of soluble methionine and S-methylmethionine. *Plant Physiol* 2002, 128, (1), 95-107.

[24] Ravanel, S.; Gakiere, B.; Job, D.; Douce, R., The specific features of methionine biosynthesis and metabolism in plants. *Proc Natl Acad Sci U S A* 1998, 95, (13), 7805-12.

[25] Ravanel, S.; Droux, M.; Douce, R., Methionine biosynthesis in higher plants. I. Purification and characterization of cystathionine gamma-synthase from spinach chloroplasts. *Arch Biochem Biophys* 1995, 316, (1), 572-84.

[26] Inaba, K.; Fujiwara, T.; Hayashi, H.; Chino, M.; Komeda, Y.; Naito, S., Isolation of an Arabidopsis thaliana Mutant, mto1, That Overaccumulates Soluble Methionine (Temporal and Spatial Patterns of Soluble Methionine Accumulation). *Plant Physiol* 1994, 104, (3), 881-887.

[27] Chiba, Y.; Ishikawa, M.; Kijima, F.; Tyson, R. H.; Kim, J.; Yamamoto, A.; Nambara, E.; Leustek, T.; Wallsgrove, R. M.; Naito, S., Evidence for autoregulation of cystathionine gamma-synthase mRNA stability in Arabidopsis. *Science* 1999, 286, (5443), 1371-4.

[28] Chiba, Y.; Sakurai, R.; Yoshino, M.; Ominato, K.; Ishikawa, M.; Onouchi, H.; Naito, S., S-adenosyl-L-methionine is an effector in the posttranscriptional autoregulation of the cystathionine gamma-synthase gene in Arabidopsis. *Proc Natl Acad Sci U S A* 2003, 100, (18), 10225-30.

[29] Onouchi, H.; Lambein, I.; Sakurai, R.; Suzuki, A.; Chiba, Y.; Naito, S., Autoregulation of the gene for cystathionine gamma-synthase in Arabidopsis: post-transcriptional regulation induced by S-adenosylmethionine. *Biochem Soc Trans* 2004, 32, (Pt 4), 597-600.

[30] Onouchi, H.; Nagami, Y.; Haraguchi, Y.; Nakamoto, M.; Nishimura, Y.; Sakurai, R.; Nagao, N.; Kawasaki, D.; Kadokura, Y.; Naito, S., Nascent peptide-mediated translation elongation arrest coupled with mRNA degradation in the CGS1 gene of Arabidopsis. *Genes Dev* 2005, 19, (15), 1799-810.

[31] Murota, K.; Hagiwara-Komoda, Y.; Komoda, K.; Onouchi, H.; Ishikawa, M.; Naito, S., Arabidopsis cell-free extract, ACE, a new in vitro translation system derived from Arabidopsis callus cultures. *Plant Cell Physiol* 2011, 52, (8), 1443-53.

[32] Hacham, Y.; Avraham, T.; Amir, R., The N-terminal region of Arabidopsis cystathionine gamma-synthase plays an important regulatory role in methionine metabolism. *Plant Physiol* 2002, 128, (2), 454-62.

[33] Hacham, Y.; Schuster, G.; Amir, R., An in vivo internal deletion in the N-terminus region of Arabidopsis cystathionine gamma-synthase results in CGS expression that is insensitive to methionine. *Plant J* 2006, 45, (6), 955-67.

[34] Di, R.; Kim, J.; Martin, M. N.; Leustek, T.; Jhoo, J.; Ho, C. T.; Tumer, N. E., Enhancement of the primary flavor compound methional in potato by increasing the level of soluble methionine. *J Agric Food Chem* 2003, 51, (19), 5695-702.

[35] Demidov, D.; Horstmann, C.; Meixner, M.; Pickard, T.; Saalbach, I.; Galili, G.; Muntz, K., Additive effects of the feed-back insensitive bacterial aspartate kinase and the Brazil nut 2S albumin on the methionine content of transgneic narbon bean (*Vicia narbonensis* L.). *Mol Breed* 2003, 11, 187-201.
[36] Amir, R.; G., G., Approaches to improve the nutritional values of transgenic plants by increasing their methionine content. In *Advances in plant physiology*, Hemantaranjan, A., Ed. Scientific publishers: Jodhpur, India, 2003; Vol. 6, pp 61-77.
[37] Avraham, T.; Badani, H.; Galili, S.; Amir, R., Enhanced levels of methionine and cysteine in transgenic alfalfa (Medicago sativa L.) plants over-expressing the Arabidopsis cystathionine gamma-synthase gene. *Plant Biotechnol J* 2005, 3, (1), 71-9.
[38] Katz, Y. S.; Galili, G.; Amir, R., Regulatory role of cystathionine-gamma-synthase and de novo synthesis of methionine in ethylene production during tomato fruit ripening. *Plant Mol Biol* 2006, 61, (1-2), 255-68.
[39] Hesse, H.; Kreft, O.; Maimann, S.; Zeh, M.; Hoefgen, R., Current understanding of the regulation of methionine biosynthesis in plants. *J Exp Bot* 2004, 55, (404), 1799-808.
[40] Kreft, O.; Hoefgen, R.; Hesse, H., Functional analysis of cystathionine gamma-synthase in genetically engineered potato plants. *Plant Physiol* 2003, 131, (4), 1843-54.
[41] Hacham, Y.; Matityahu, I.; Schuster, G.; Amir, R., Overexpression of mutated forms of aspartate kinase and cystathionine gamma-synthase in tobacco leaves resulted in the high accumulation of methionine and threonine. *Plant J* 2008, 54, (2), 260-71.
[42] Dancs, G.; Kondrak, M.; Banfalvi, Z., The effects of enhanced methionine synthesis on amino acid and anthocyanin content of potato tubers. *BMC Plant Biol* 2008, 8, 65.
[43] Loizeau, K.; Gambonnet, B.; Zhang, G. F.; Curien, G.; Jabrin, S.; Van Der Straeten, D.; Lambert, W. E.; Rebeille, F.; Ravanel, S., Regulation of one-carbon metabolism in Arabidopsis: the N-terminal regulatory domain of cystathionine gamma-synthase is cleaved in response to folate starvation. *Plant Physiol* 2007, 145, (2), 491-503.
[44] Avraham, T.; Amir, R., The expression level of threonine synthase and cystathionine-gamma-synthase is influenced by the level of both threonine and methionine in Arabidopsis plants. *Transgenic Res* 2005, 14, (3), 299-311.
[45] Riedel, K.; Mangelsdorf, C.; Streber, W.; Willmitzer, L.; Hoefgen, R.; Hesse, H., Cloning and characterization of cystathionine gamma synthase from *Solanum tuberosum* L. *Plant Biol* 1999, 1, 638-644.
[46] Gakiere, B.; Ravanel, S.; Droux, M.; Douce, R.; Job, D., Mechanisms to account for maintenance of the soluble methionine pool in transgenic Arabidopsis plants expressing antisense cystathionine gamma-synthase cDNA. *C R Acad Sci III* 2000, 323, (10), 841-51.
[47] Droux, M. G., B. Denis, L. Ravanel, Tabe, L. Lappartient, A.G. Job, D., *Methionine biosynthesis in plants: Biochemical and regulatory aspects*. Paul Haupt: Bern, Switzerland, 2000; p 73-92.
[48] Supriyadi, S. M., Wu S, Tomita N, Fujita A, Watanabe N., Biogenesis of volatile methyl esters in snake fruit (*Salacca edulis*, Reinw) cv. Pondoh. *Biosci Biotechnol Biochem.* 2003, 67, 1267-1271.
[49] Amir, R.; Hacham, Y.; Galili, G., Cystathionine gamma-synthase and threonine synthase operate in concert to regulate carbon flow towards methionine in plants. *Trends Plant Sci* 2002, 7, (4), 153-6.

[50] Zeh, M.; Casazza, A. P.; Kreft, O.; Roessner, U.; Bieberich, K.; Willmitzer, L.; Hoefgen, R.; Hesse, H., Antisense inhibition of threonine synthase leads to high methionine content in transgenic potato plants. *Plant Physiol* 2001, 127, (3), 792-802.

[51] Bartlem, D. L., I. Okamoto, T. Itaya, A. Uda, Y. Kijima, Tamaki, Y. Nambara, E. Naito, S., Mutation in the threonine synthase gene results in an over-accumulation of soluble methionine in Arabidopsis. *Plant Physiol* 2000, 123, 101-10.

[52] Curien, G.; Job, D.; Douce, R.; Dumas, R., Allosteric activation of Arabidopsis threonine synthase by S-adenosylmethionine. *Biochemistry* 1998, 37, (38), 13212-21.

[53] Aubert, S.; Curien, G.; Bligny, R.; Gout, E.; Douce, R., Transport, compartmentation, and metabolism of homoserine in higher plant cells. *Plant Physiology* 1998, 116, 547-557.

[54] Shen, B.; Li, C.; Tarczynski, M. C., High free-methionine and decreased lignin content result from a mutation in the Arabidopsis S-adenosyl-L-methionine synthetase 3 gene. *Plant J* 2002, 29, (3), 371-80.

[55] Lee, M.; Martin, M. N.; Hudson, A. O.; Lee, J.; Muhitch, M. J.; Leustek, T., Methionine and threonine synthesis are limited by homoserine availability and not the activity of homoserine kinase in Arabidopsis thaliana. *Plant J* 2005, 41, (5), 685-96.

[56] Rinder, J.; Casazza, A. P.; Hoefgen, R.; Hesse, H., Regulation of aspartate-derived amino acid homeostasis in potato plants (Solanum tuberosum L.) by expression of E. coli homoserine kinase. *Amino Acids* 2008, 34, (2), 213-22.

[57] Shaul, O.; Galili, G., Threonine Overproduction in Transgenic Tobacco Plants Expressing a Mutant Desensitized Aspartate Kinase of Escherichia coli. *Plant Physiol* 1992, 100, (3), 1157-63.

[58] Perl, A.; Shaul, O.; Galili, G., Regulation of lysine biosynthesis in transgenic potato plants expressing a abacterial dihydrodipicolinicolinate synthase in their chloroplsts. *Plant Mol. Biol* 1992, 19, 815-823.

[59] Ben-Tzvi Tzchori, I.; Perl, A.; Galili, G., Lysine and threonine metabolism are subject to complex patterns of regulation in Arabidopsis. *Plant Mol Biol* 1996, 32, (4), 727-34.

[60] Galili, S.; Guenoune, D.; Wininger, S.; Hana, B.; Schupper, A.; Ben-Dor, B.; Kapulnik, Y., Enhanced levels of free and protein-bound threonine in transgenic alfalfa (Medicago sativa L.) expressing a bacterial feedback-insensitive aspartate kinase gene. *Transgenic Res* 2000, 9, (2), 137-44.

[61] Matityahu, I.; Kachan, L.; Bar Ilan, I.; Amir, R., Transgenic tobacco plants overexpressing the Met25 gene of Saccharomyces cerevisiae exhibit enhanced levels of cysteine and glutathione and increased tolerance to oxidative stress. *Amino Acids* 2006, 30, (2), 185-94.

[62] Czuj, T.; Żuk, M.; Starzycki, M.; Amir, R.; Szopa, J., Engineering increases in sulfur amino acid contents in flax by overexpressing the yeast Met25 gene. *Plant Science* 2009, 177, 584–592.

[63] Saito, K.; Kurosawa, M.; Tatsuguchi, K.; Takagi, Y.; Murakoshi, I., Modulation of cysteine biosynthesis in chloroplasts of transgenic tobacco overexpressing cysteine synthase [O-acetylserine(thiol)-lyase]. *Plant Physiol* 1994, 106, (3), 887-95.

[64] Blaszczyk, A.; Brodzik, R.; Sirko, A., Increased resistance to oxidative stress in transgenic tobacco plants overexpressing bacterial serine acetyltransferase. *Plant J* 1999, 20, (2), 237-43.

[65] Sirko, A.; Blaszczyk, A.; Liszewska, F., Overproduction of SAT and/or OASTL in transgenic plants: a survey of effects. *J Exp Bot* 2004, 55, (404), 1881-8.

[66] Noji, M.; Saito, M.; Nakamura, M.; Aono, M.; Saji, H.; Saito, K., Cysteine synthase overexpression in tobacco confers tolerance to sulfur-containing environmental pollutants. *Plant Physiol* 2001, 126, (3), 973-80.

[67] Stiller, I.; Dancs, G.; Hesse, H.; Hoefgen, R.; Banfalvi, Z., Improving the nutritive value of tubers: elevation of cysteine and glutathione contents in the potato cultivar White Lady by marker-free transformation. *J Biotechnol* 2007, 128, (2), 335-43.

[68] Boerjan, W.; Bauw, G.; Van Montagu, M.; Inze, D., Distinct phenotypes generated by overexpression and suppression of S-adenosyl-L-methionine synthetase reveal developmental patterns of gene silencing in tobacco. *Plant Cell* 1994, 6, (10), 1401-14.

[69] Nikiforova, V. J.; Bielecka, M.; Gakiere, B.; Krueger, S.; Rinder, J.; Kempa, S.; Morcuende, R.; Scheible, W. R.; Hesse, H.; Hoefgen, R., Effect of sulfur availability on the integrity of amino acid biosynthesis in plants. *Amino Acids* 2006, 30, (2), 173-83.

[70] Nikiforova, V. J.; Kopka, J.; Tolstikov, V.; Fiehn, O.; Hopkins, L.; Hawkesford, M. J.; Hesse, H.; Hoefgen, R., Systems rebalancing of metabolism in response to sulfur deprivation, as revealed by metabolome analysis of Arabidopsis plants. *Plant Physiol* 2005, 138, (1), 304-18.

[71] Negrutiu, I. D. B., D. Dirks, R. Jacobs, M., Amino acid auxotrophs from protoplast culturees of *Nicotiana plumbagiifolia*, Viviani. *Mol. Gen. Genet.* 1985, 199, 330-337.

[72] Maimann, S.; Wagner, C.; Kreft, O.; Zeh, M.; Willmitzer, L.; Hofgen, R.; Hesse, H., Transgenic potato plants reveal the indispensable role of cystathionine beta-lyase in plant growth and development. *Plant J* 2000, 23, (6), 747-58.

[73] Hacham, Y.; Gophna, U.; Amir, R., In vivo analysis of various substrates utilized by cystathionine gamma-synthase and O-acetylhomoserine sulfhydrylase in methionine biosynthesis. *Mol Biol Evol* 2003, 20, (9), 1513-20.

[74] Maimann, S.; Hoefgen, R.; Hesse, H., Enhanced cystathionine beta-lyase activity in transgenic potato plants does not force metabolite flow towards methionine. *Planta* 2001, 214, (2), 163-70.

[75] Gong, P.; Zhang, J.; Li, H.; Yang, C.; Zhang, C.; Zhang, X.; Khurram, Z.; Zhang, Y.; Wang, T.; Fei, Z.; Ye, Z., Transcriptional profiles of drought-responsive genes in modulating transcription signal transduction, and biochemical pathways in tomato. *J Exp Bot* 2010, 61, (13), 3563-75.

[76] Eckermann, C.; Eichel, J.; Schroder, J., Plant methionine synthase: new insights into properties and expression. *Biol Chem* 2000, 381, (8), 695-703.

[77] Ravanel, S.; Block, M. A.; Rippert, P.; Jabrin, S.; Curien, G.; Rebeille, F.; Douce, R., Methionine metabolism in plants: chloroplasts are autonomous for de novo methionine synthesis and can import S-adenosylmethionine from the cytosol. *J Biol Chem* 2004, 279, (21), 22548-57.

[78] , M.; Leggewie, G.; Hoefgen, R.; Hesse, H., Cloning and characterization of a cDNA encoding a cobalamin-independent methionine synthase from potato (Solanum tuberosum L.). *Plant Mol Biol* 2002, 48, (3), 255-65.

[79] Gallardo, K.; Job, C.; Groot, S. P.; Puype, M.; Demol, H.; Vandekerckhove, J.; Job, D., Importance of methionine biosynthesis for Arabidopsis seed germination and seedling growth. *Physiol Plant* 2002, 116, (2), 238-247.

[80] Zegouti, H.; Jones, B.; Frasse, P.; Marty, C.; Maitre, B.; Latche, A.; Pech, J. C.; M., B., Ethylene-regulated gene expression in tomato fruits: characterization of novel ethylene responsive and ripening related genes isolated by differential display. *Plant J.* 1999, 18, 589-600.

[81] Azevedo, R. A.; Lancien, M.; Lea, P. J., The aspartic acid metabolic pathway, an exciting and essential pathway in plants. *Amino Acids* 2006, 30, (2), 143-62.

[82] Shaul, O.; Galili, G., Concerted regulation of lysine and threonine synthesis in tobacco plants expressing bacterial feedback-insensitive aspartate kinase and dihydrodipicolinate synthase. *Plant Mol Biol* 1993, 23, (4), 759-68.

[83] Perl, A.; Galili, S.; Shaul, O.; Ben Tzvi-Tzchori, I.; Galili, G., Bacterial dihydrodipicolinicolinate synthase and desedsitized aspartate kinase: two novel selectable markers for plant transformation. *Bio/Technol* 1993, 11:715-718.

[84] Hacham, Y.; Song, L.; Schuster, G.; Amir, R., Lysine enhances methionine content by modulating the expression of S-adenosylmethionine synthase. *Plant J* 2007, 51, (5), 850-61.

[85] Brinch-Pedersen, H.; Galili, G.; Knudsen, S.; Holm, P. B., Engineering of the aspartate family biosynthetic pathway in barley (Hordeum vulgare L.) by transformation with heterologous genes encoding feed-back-insensitive aspartate kinase and dihydrodipicolinate synthase. *Plant Mol Biol* 1996, 32, (4), 611-20.

[86] Zhu, X.; Galili, G., Increased lysine synthesis coupled with a knockout of its catabolism synergistically boosts lysine content and also transregulates the metabolism of other amino acids in Arabidopsis seeds. *Plant Cell* 2003, 15, (4), 845-53.

[87] Zhu, X.; Galili, G., Lysine metabolism is concurrently regulated by synthesis and catabolism in both reproductive and vegetative tissues. *Plant Physiol* 2004, 135, (1), 129-36.

[88] Pimenta, M. K., T. Larondelle, Y. Dohmae, N. Kamiya, Y., *S*-adenosyl-L-methionine:L-methionine S-methyltransferase from germinating barley. Purification and localization. *Plant Physiol.* 1998, 118, 431-438.

[89] Rebeille, F.; Jabrin, S.; Bligny, R.; Loizeau, K.; Gambonnet, B.; Van Wilder, V.; Douce, R.; Ravanel, S., Methionine catabolism in Arabidopsis cells is initiated by a gamma-cleavage process and leads to S-methylcysteine and isoleucine syntheses. *Proc Natl Acad Sci U S A* 2006, 103, (42), 15687-92.

[90] Bourgis, F.; Roje, S.; Nuccio, M. L.; Fisher, D. B.; Tarczynski, M. C.; Li, C.; Herschbach, C.; Rennenberg, H.; Pimenta, M. J.; Shen, T. L.; Gage, D. A.; Hanson, A. D., S-methylmethionine plays a major role in phloem sulfur transport and is synthesized by a novel type of methyltransferase. *Plant Cell* 1999, 11, (8), 1485-98.

[91] Lee, M.; Huang, T.; Toro-Ramos, T.; Fraga, M.; Last, R. L.; Jander, G., Reduced activity of Arabidopsis thaliana HMT2, a methionine biosynthetic enzyme, increases seed methionine content. *Plant J* 2008, 54, (2), 310-20.

[92] Gallardo, K.; Firnhaber, C.; Zuber, H.; Hericher, D.; Belghazi, M.; Henry, C.; Kuster, H.; Thompson, R., A combined proteome and transcriptome analysis of developing Medicago truncatula seeds: evidence for metabolic specialization of maternal and filial tissues. *Mol Cell Proteomics* 2007, 6, (12), 2165-79.

[93] Karchi, H.; Shaul, O.; Galili, G., Seed specific expression of a bacterial desensitized aspartate kinase increases the production of seed threonine and methionine in transgenic tobacco. *Plant J.* 1993, 3, 721-727.

[94] Burstenbinder, K.; Rzewuski, G.; Wirtz, M.; Hell, R.; Sauter, M., The role of methionine recycling for ethylene synthesis in Arabidopsis. *Plant J* 2007, 49, (2), 238-49.

[95] Giovanelli, J. G., Sulfur amino acids in plants: an overview. *Methods in Enzymol.* 1987, 143, 419-428.

[96] Kumar, A.; Taylor, M. A.; Mad Arif, S. A.; Davies, H. V., Potato plants expressing antisense and sense S-adenosylmethionine decarboxylase (SAMDC) transgenes show altered levels of polyamines and ethylene: antisense plants dispaly abnormal phenotypes. *Plant J.* 1996, 9, 147-158.

[97] Goto, D. B.; Ogi, M.; Kijima, F.; Kumagai, T.; van Werven, F.; Onouchi, H.; Naito, S., A single-nucleotide mutation in a gene encoding S-adenosylmethionine synthetase is associated with methionine over-accumulation phenotype in Arabidopsis thaliana. *Genes Genet Syst* 2002, 77, (2), 89-95.

[98] Goyer, A.; Collakova, E.; Shachar-Hill, Y.; Hanson, A. D., Functional characterization of a methionine gamma-lyase in Arabidopsis and its implication in an alternative to the reverse trans-sulfuration pathway. *Plant Cell Physiol* 2007, 48, (2), 232-42.

[99] Saint-Girons, I. P., C. Zakin, M.M. Barzu, O. Cohen, G. N., Methionine biosynthesis in enterobacteriaceae: biochimical, regulatory, and evolutionary aspects. *CRC Crit Rev Biochem* 1988, 23 Suppl 1, S1-S42.

[100] Joshi, V.; Jander, G., Arabidopsis methionine gamma-lyase is regulated according to isoleucine biosynthesis needs but plays a subordinate role to threonine deaminase. *Plant Physiol* 2009, 151, (1), 367-78.

[101] Less, H.; Galili, G., Coordinations between gene modules control the operation of plant amino acid metabolic networks. *BMC Syst Biol* 2009, 3, 14.

[102] Nour-Eldin, H. H.; Andersen, T. G.; Burow, M.; Madsen, S. R.; Jorgensen, M. E.; Olsen, C. E.; Dreyer, I.; Hedrich, R.; Geiger, D.; Halkier, B. A., NRT/PTR transporters are essential for translocation of glucosinolate defence compounds to seeds. *Nature* 2012, 488, (7412), 531-4.

[103] Sonderby, I. E.; Geu-Flores, F.; Halkier, B. A., Biosynthesis of glucosinolates--gene discovery and beyond. *Trends Plant Sci* 2010, 15, (5), 283-90.

[104] Sawada, Y.; Kuwahara, A.; Nagano, M.; Narisawa, T.; Sakata, A.; Saito, K.; Hirai, M. Y., Omics-based approaches to methionine side chain elongation in Arabidopsis: characterization of the genes encoding methylthioalkylmalate isomerase and methylthioalkylmalate dehydrogenase. *Plant Cell Physiol* 2009, 50, (7), 1181-90.

[105] Saito, K., Regulation of sulfate transport and synthesis of sulfur-containing amino acids. *Curr Opin Plant Biol* 2000, 3, (3), 188-95.

[106] Jez, J. M., Plant sulfur compounds and human helath. In *Sulfur: a missing link between soils, crops and nutrition.* , Jez, J. M., Ed. ASA-CSSA-SSSA: Madison, WI, USA, 2008; pp 281-291.

[107] Galili, G.; Amir, R.; Hoefgen, R.; Hesse, H., Improving the levels of essential amino acids and sulfur metabolites in plants. *Biol Chem* 2005, 386, (9), 817-31.

[108] Waterlow, J. C.; Payne, D. R., The protein gap. *Nature* 1975, 258, 113-117.

[109] Fukagawa, N. K., Sparing of methionine requirements: evaluation of human data takes sulfur amino acids beyond protein. *J Nutr.* 2006, 136, 1676S-1681S.

[110] Fukagawa, N. K.; Galbraith, R. A., Advancing age and other factors influencing the balance between amino acid requirements and toxicity. *J Nutr.* 2004, 134, 1569S-1574S.

[111] Poirier, L., The effects of diet, genetics and chemicals on toxicity and aberrant DNA methylation: an introduction. *J Nutr.* 2002, 132, 2336S-2339S.

[112] Lertratanangkoon, K.; Orkiszewski, R. S.; Scimeca, J. M., Methyl-donor deficiency due to chemically induced glutathione depletion. . *Cancer Res.* 1996, 56, 995–1005.

[113] Minh, D. V.; Ogle, B., Effect of Scavenging and Supplementation of Lysine and Methionine on the Feed Intake, Performance and Carcase Quality of Improved Dual-Purpose Growing Chickens *Tropical animal health and production* 2005, 37, 573-587.

[114] Sadar, P.; Abid, M.; Randhawa, H. S.; Prabhakar, H. S., Effect of dietary lysine and methionine supplementation on growth, nutrient utilization, carcass compositions and haemato-biochemical status in Indian Major Carp, Rohu (Labeo rohita H.) fed soy protein-based diet. *Aquaculture Nutrition* 2009, 15, 339-346.

[115] Xu, S.; Harrison, J. H.; Chalupa, W.; Sniffen, C.; Julien, W.; Sato, H.; Fujieda, T.; Watanabe, K.; Ueda, T.; Suzuki, H., The effect of ruminal bypass lysine and methionine on milk yield and composition of lactating cows. *Dairy Sci* 1998, 81, 1062-77.

[116] Deikman, J.; Xu, R.; Kneissl, M.; Ciardi, J.; Kim, K.; Pelah, D., Separation of cis elements responsive to ethylene, fruit development, and ripening in the 5'-flanking region of the ripening-related E8 gene. *Plant Mol Biol* 1998, 37, 1001-1011.

[117] Imsande, J., Selection of soybean mutants with increased concentrations of seed methionine and cysteine. *Crop Science* 2001, 41, 510-515.

[118] Amir, R.; Tabe, L., Molecular approaches to improving plant methionine content. In *Plant Genetic Engineering Vol 8: Metabolic engineering and molecular farming II.* , P.S., P. K. J. J. a. R., Ed. Studium Press LLC: Huston, Texas, 2006; pp 1-26.

[119] Tabe, M. D. M., Limits to sulfur accumulation in transgenic Lupin seeds expressing a foreign sulfur-rich protein. *Plant Physiol* 2002, 128, 1137-1148.

[120] Bagga, S.; Ponteza, C.; Ross, J.; Martin, M. N.; Luestek, T.; Sengupta-Gopalan, C., A transgene for high methionine protein is post transcriptionally regulated by methionine *In vitro Cell Dev Biol Plant* 2005, 41, 731-741.

[121] Golan, A.; Matitiyho, I.; Avraham, T.; Badani, H.; Galili, S.; Amir, R., Soluble methionine enhances accumulation of a 15 kDa zein, a methionine-rich storage protein, in transgenic alfalfa but not in transgenic tobacco plants. *J Exp Bot* 2005, 56, (419), 2443-52.

In: Methionine
Editor: Alexander Snegursky

ISBN: 978-1-62618-246-2

Chapter V

Methionine: Electronic Structure

Oksana Plekan*[1]*, Vitaliy Feyer*[1,2]*, Robert Richter*[1]*,
***Marcello Coreno*[3]*, Monica de Simone*[4]*, Kevin C. Prince**[*1,4]
***and Vincenzo Carravetta*[5]
[1]Sincrotrone Trieste S.C.p.A., in Area Science Park, Trieste, Italy
[2]Forschungszentrum Jülich GmbH, Peter Grünberg Institute, Electronic Properties (PGI-6), Leo-Brandt-Straße, Jülich, Germany
[3]CNR-IMIP, Montelibretti (Rome), Italy
[4]IOM-CNR, in Area Science Park, Basovizza, Trieste, Italy
[5]CNR-IPCF, Pisa, Italy

Abstract

We review the electronic structure of methionine, determined by means of photoemission and photoabsorption spectroscopy. Valence band spectroscopy yields the energies of the molecular orbitals responsible for the chemical bonding within the molecule. The Highest Occupied Molecular Orbital of methionine has sulfur lone pair character, and we show how this helps to stabilize the molecule against photofragmentation, compared for example with amino acids having aliphatic side chains. The core level spectra give information about the local chemical properties of each heavy atom in the molecule, while x-ray absorption sheds light on the unoccupied molecular orbitals. The two lowest orbitals accessible to carbon 1s excitation both have partial sulfur character. However neither of these is the Lowest Unoccupied Molecular orbital, which is instead the antibonding π orbital of the carboxylic acid group.

[*] Email: Prince@Elettra.Trieste.It

Introduction

Nowadays theoretical methods are very advanced and it is possible to calculate the properties of increasingly complex molecular systems, and predict their chemical and physical behavior. The gas phase spectra of free molecules can then be used as benchmarks to check the quality of ab initio calculations of free molecules, in which the intrinsic properties of the molecules are predicted. In addition, gas phase spectra can be compared with spectra in liquids, solutions and matrix isolation, in order to assess the interaction with the solvent and neighboring molecules [1-5].

Among the many organic molecules of biological relevance, amino acids are of considerable importance as building blocks of peptides and proteins, but also for their participation in a number of physiological processes. As a consequence, gas-phase spectroscopy of the 20 naturally occurring amino acids has attracted the interest of researchers for several decades. There has been a growing interest in characterization of biologically relevant systems by synchrotron radiation, because core level spectroscopy can provide insight into molecular electronic structure, and synchrotron radiation is an excellent source for photoemission and photoabsorption, particularly near edge X-ray absorption fine structure (NEXAFS) spectroscopy. NEXAFS spectra of amino acids have been reported of experimental studies in different phases, such as adsorbed on surfaces, dissolved in liquids and in the gas phase [6-13] and predicted by ab-initio calculations for the isolated molecules[14-16].

Methionine (Met) is one of the essential amino acids (see Figure 1) and it must be present in the diet of mammals to ensure normal growth [17]. As one of the two sulfur-containing amino acids, Met has attracted growing interest recently since it is thought to play an important role in the severe nephrotoxicity of cisplatin, inactivation of Pt complexes, and tumor resistance to treatment with Pt drugs [18]. Like many other amino acids, Met al.so has several possible ionic structural forms, depending on its local environment: in the crystalline state or in aqueous solutions at normal pH, it has zwitterionic form, in which the molecule contains both a positive and a negative charge, while the isolated molecule is in the un-ionized form. In acidic or basic solution, it may also be present as a cation or an anion, respectively. Literature reports of spectroscopic studies and ab initio calculations of the electronic structure of Met are very limited. Recently the gas phase Fourier transform infrared (FTIR) spectra of methionine and other natural amino acids have been measured and compared to density functional calculations at the B3LYP/6-31G(d,p) level of theory [19]. Sulfur K-edge NEXAFS has been measured and calculated for DL-methionine in order to understand the activity of proteins due to the modification of sulfur groups in the sulfur-containing amino acids [20]. Proton affinity and protonation entropy of Met have been determined by the extended kinetic method from ESI-Q-TOF tandem mass spectrometry experiments [21].

The present work reviews the high-resolution outer and inner valence band (VB) spectra, as well as core level X-ray photoemission (XPS) and NEXAFS studies at the carbon, nitrogen, and oxygen K-edges of Met. All data have been obtained in the gas phase using synchrotron light as the source. We discuss the spectral features characteristic of the chemical bonds in the carbon, nitrogen, and oxygen XPS and NEXAFS spectra of methionine in comparison with the smallest of the proteinogenic amino acids, glycine (Gly). In addition, the photoionization mass spectra of methionine at five different photon energies are presented in order to get important information about, and understanding of, radiation damage in biological systems.

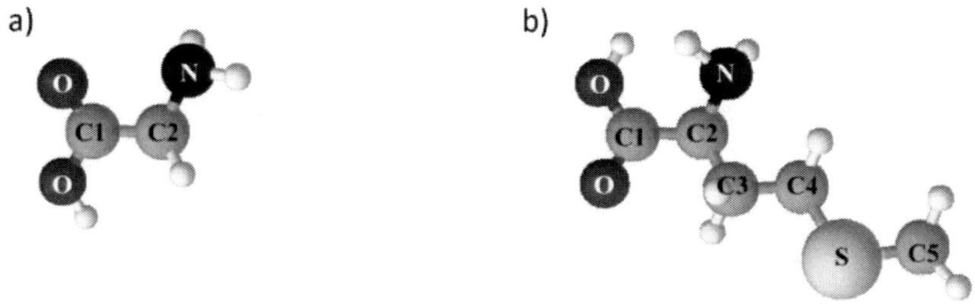

Figure 1. Schematic structures of (a) glycine and (b) methionine. One of the hydrogen atoms attached to the α carbon atom C2 in glycine is replaced in methionine by the C3-C4-S-C5 side chain.

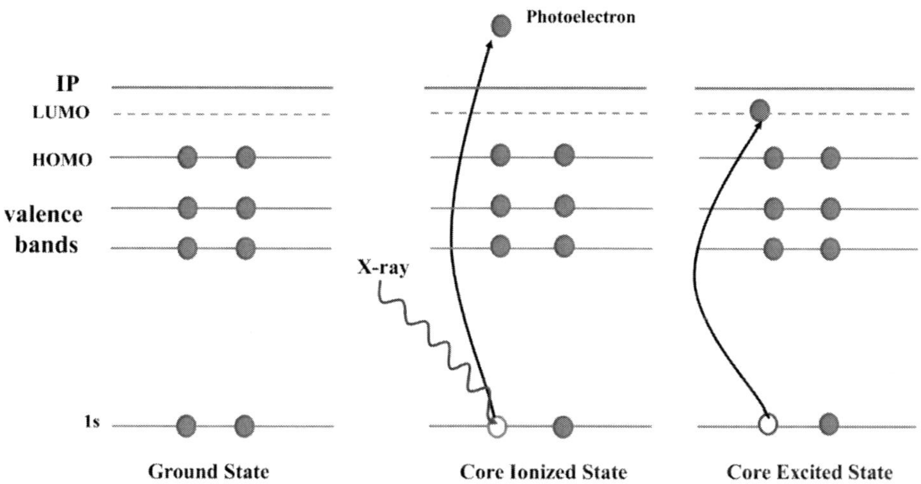

Figure 2. Sketch of electronic structure, photoionization and photoabsorption. The vertical direction indicates energy, not to scale, and the valence band includes the HOMO (Highest Occupied Molecular Orbital). The LUMO (Lowest Unoccupied Molecular Orbital) lies between the HOMO and the IP (Ionization Potential). Centre panel: schematic view of the photoemission process. Right panel: schematic photoabsorption process.

The reader may be unfamiliar with the main methods used in this study, namely photoionization, x-ray photoabsorption and photoionization mass spectrometry, so we here give a brief description. The first technique we have applied is core and valence ionization, Fig. 2. In this method the molecule under study absorbs a photon, an electron from the core or valence region is ejected, and the kinetic energy of this electron is measured. This is the photoelectric effect, which Albert Einstein explained and for which he won the Nobel prize. The energy of the photon is quantized, so it is well-defined, and the difference between the photon energy and the kinetic energy of the electron is defined as the Binding Energy. If the Schrödinger equation for the electrons of the molecule is solved in the independent particle approximation, eigenfunctions are found for core, or strongly bound, electrons and for valence electrons, and the latter eigenfunctions constitute the molecular orbitals which give rise to chemical bonding. Within an independent electron picture of molecules, the Binding Energy can be identified with the energy of a particular atomic or molecular orbital: this approximation is known as Koopman's theorem.

The independent electron picture is a rather basic description and it does not take account of effects such as correlation, that is to say, that one electron responds, by the electric field, to the instantaneous positions of all the other electrons and not just to an average. The orbitals provide an approximate description of the electronic structure, that for instance is known to break down for the inner valence shells. Another effect important in photoionization is electronic relaxation, that is neglected by the Koopman's approximation and that describes the response of the other electrons to the sudden creation of a hole when the photoelectron is removed. To put it another way, the orbitals of the electrons in the neutral molecule are not the same as the orbitals of the singly charged ion created by photoionization. It is intuitively clear that it must be so: the charge of the system has changed, so the electrons adjust in response. The shift in binding energy due to electron relaxation is limited for valence ionization because valence orbitals are diffuse, so that the loss of a valence electron does not significantly affect the inner ones. Also, the Koopman's approximation sometimes provides a good enough prediction for valence ionization because the neglected effect of electron relaxation is opposite in sign to that of electron correlation and similar in size.

The core electrons in an atom or molecule are not delocalized, but instead are confined to a small volume near the nucleus of the atom from which they originate. This has the advantage that their energies represent a very local probe of the chemical state of that atom in the molecule, and for this discovery, Kai Siegbahn won the Nobel prize. In the simplest view of core ionization, the binding energy of the outgoing photoelectron is higher if the atom is bonded to electronegative neighbors, because these atoms withdraw charge (increasing the local Madelung potential), and lower if the neighbors are electropositive. This is the ESCA (Electron Spectroscopy for Chemical Analysis) model, and it is valid for a wide range of materials. However once again, a more sophisticated model takes into account relaxation, the response of the valence electrons to the sudden creation of a core hole. The outer electrons react (that is, their orbitals contract) in response to the creation of a core hole, and electrically screen the hole. Since the core hole is highly localized, the effect of relaxation is stronger than for valence levels. Such effects are often called "final state effects", because they relate to the energy of the ionized system, whereas the ESCA model relates mostly to initial state effects, that is, the chemical state in the neutral ground state. Often the final state effects are uniform, that is, all core levels of atoms of the same kind have a similar relaxation energy, but in some cases there can be significant differences.

The second spectroscopic method which we use here is soft x-ray photoabsorption, Fig. 2. In this technique, an electron is excited from the core to an unoccupied orbital, and thus probes the empty states, such as the LUMO (Lowest Unoccupied Molecular Orbital) and other antibonding states. Experimentally, the photon energy is scanned and a signal proportional to the absorption cross-section is measured, for example the total number of ions or electrons produced. When absorption is strong (a resonance) the yield, and therefore the cross-section, is high. Like core photoemission, it is a local probe and determines the density of unoccupied orbitals at the site of the core excitation.

The third method is photoionization mass spectroscopy, in which molecules are ionized with low energy photons, sufficient to remove a valence electron. Photoionization mass spectroscopy is a broad field, and is often performed close to threshold to minimize fragmentation, and to determine the first ionization potential (IP) of the molecule and ion appearance potentials, that is, the energy which it is necessary to put into a molecule to create a particular ion. The reason that molecules may fragment on ionization is that excess energy may be present in the ion which needs to be dissipated. For instance, the minimum energy geometry of the ion may be different from the neutral state. Thus when the ion is created it has the bond lengths and bond angles of the ground state, and so is unstable. Even if both geometries are similar, the ion may be created in an electronically excited state. The processes available for dissipating energy are nuclear motion, including bond breaking and rearrangement, fluorescence and, for inner electron ionization, non radiative (Auger) decay.

A more common method of ionizing molecules to measure mass spectra is electron impact ionization, and with respect to this method, photoionization is a more specific excitation: electron impact can excite dipole allowed ionic states, but also non-dipole states, which may have different decay modes from dipole excited states. Nevertheless, even after photoionization the decay is complex: different ionic states may decay in different ways, vibrational energy plays a role, and cross-sections for photoionization change with energy. The information provided is important for understanding radiation damage in biological systems.

Experimental Section

The measurements were performed at the gas-phase photoemission beamline, Elettra, Trieste [22] and the experimental methods have been described in detail elsewhere [23]. The methionine was obtained from Sigma–Aldrich in the form of crystalline powder with minimum purity of 99% and used without any further purification. A few grams of powder were placed in a noncommercial, non-inductively wound furnace equipped with a chromel/alumel thermocouple. The evaporation temperature was 445 K and the pressure in the chamber remained in the 10^{-8} mbar range during the experiments.

It is important to note that biomolecules, like amino acids and peptides, have very low vapor pressures and are well known for their thermal instability. Therefore before the experiment with synchrotron radiation, the photoionisation mass spectra of methionine were taken to check for purity and absence of thermal decomposition and determine the appropriate temperature for evaporation [24].

Photoionization mass spectra of methionine were measured with a Wiley-McLaren type time of flight (TOF) spectrometer, built in-house, with a mass resolving power of 200. A noble gas resonance lamp (Specs model UVS 300A) was used to provide radiation at 21.2 (He I), 16.67 (Ne I), 11.62 (Ar I), 10.0 (Kr I) and 8.43 eV (Xe I) for ionizing the Met molecule. The mass spectra were recorded by the Photo Electron Photo Ion Coincidence (PEPICO) technique. The ions were accelerated and focused by an electric field and detected by a microchannel plate, and gave rise to stop signals sent to a time-to-digital converter (TDC), while electrons were detected without energy analysis by a channel electron multiplier and provided the start signal to the TDC. The design and operating principles of the TOF are described in detail in [24].

The valence spectra were recorded with an incident photon energy 99 eV. The S 2p, C, N and O 1s core level spectra were measured at 256 eV, 382, 495 and 628 eV photon energy respectively and referenced to published values of reference gases: S 2p of sulfur hexafluoride [25, 26], C 1s of carbon dioxide [27], N 1s of nitrogen and O 1s of carbon dioxide [28]. The spectra were taken with a total resolution (photons+analyser) of 0.45, 0.25, 0.57, 0.59 and 0.53 eV at hv=99, 256, 382, 495 and 628 eV respectively. The NEXAFS spectra were recorded by collecting the ion yield signal using a channel electron multiplier placed close to the ionization region. In the geometry used, some energetic ions may be lost, so the ion yield is partial rather than total. The spectra at the carbon, nitrogen, and oxygen K edges were normalized to the photon flux measured by a photodiode. The energy scales of the spectra were calibrated by taking simultaneous spectra of the samples and a calibrant gas introduced into the experimental chamber. The NEXAFS spectra were recorded with energy resolution of 0.07, 0.06, and 0.10 eV at the C, N, and O K edges, respectively. The energies were calibrated to the following resonances: 290.77 eV (C 1s $\rightarrow \pi$, CO2) [29], 400.87 eV (N 1s $\rightarrow \pi$, N2) [30], 535.4 eV (O 1s $\rightarrow \pi$, CO2) [31].

Results and Discussion

Valence Photoemission Spectra

The valence photoemission spectrum of methionine measured at photon energy 99 eV is shown in Figure 3. For comparison the valence spectrum of Gly is also presented in Figure 3, measured under the same experimental conditions. The energies of the peaks in the spectra are similar to those in the previously reported He I spectra [32], and there is good agreement between the peak positions; the relative peak intensities are different due to the different photon energy. It is known that amino acids in the gas phase exist as several conformers (see Figure 4) [8, 9, 33], and we follow Dehareng and Dive [33] for the assignments of the molecular orbital energies. They studied the variation of the vertical ionization energies of 18 of the 20 natural amino acids as a function of their gas phase conformation. The conformers are of two main types, and depend on the internal hydrogen bonding between the amino and carboxylic acid group. In Fig. 4(a), the conformer CF1 of Met is shown, in which there is a hydrogen bond between the NH2 (donor) and carboxy oxygen (acceptor) groups. The alternative, CF2, is shown in Fig. 4(b), in which the donor group is the carboxylic acid OH and the acceptor is the nitrogen of the amino group. Dehareng and Dive also considered conformers involving hydrogen bond interactions with the sulfur group. For CF2, these are labeled CF2(1) and CF2(2), with and without a hydrogen bond from the amino group to the sulfur atom.

In Table 1 the ionization energies of the first four ionic states for the lowest energy conformers of methionine are presented. We list only the first four frontier orbitals, as these are most important in determining the chemical interaction of methionine.

The first symmetric peak in the valence photoemission spectra of Met at BE=8.65 eV corresponds to the ionization of the S lone-pair orbital (nS). According to the calculations [33] of the vertical ionization energies of these three conformers of methionine, the conformer CF1 is predicted to have a first ionization energy of 8.09 eV. There is however no peak in the spectrum at this energy: the first peak is at 8.65 eV, consistent with the calculated energies of 8.60 and 8.58 eV for conformers CF2(1) and CF2(2). Thus, we conclude that CF1 is not significantly populated at the temperature of our experiment. Most methionine molecules adopt the conformation shown in Fig. 4(b), with a hydroxyl-amino hydrogen bond. The experimental data do not allow us to determine the relative populations of molecules with and without hydrogen bonds to sulfur.

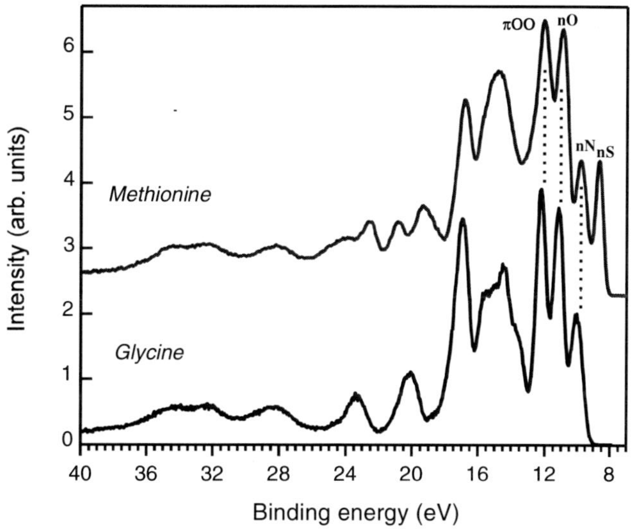

Figure 3. Valence band photoemission spectra of Met (top curve, blue) and Gly (bottom curve, black) at photon energy 99 eV.

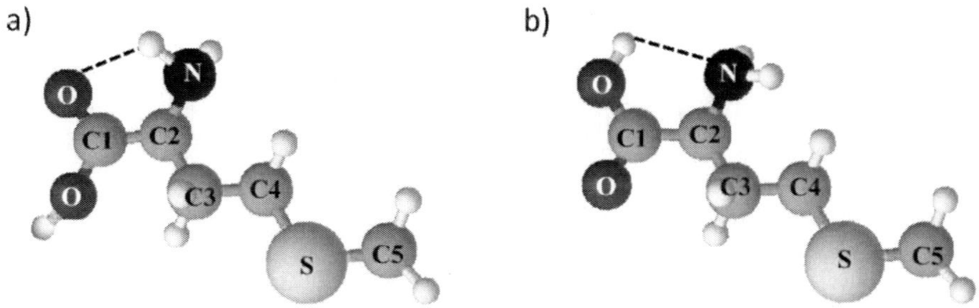

Figure 4. The two lowest energy conformers for methionine: (a) CF1 and (b) CF2.

The next bands correspond to ionization of the nitrogen lone-pair orbital (nN), BE =9.80 eV, and of an oxygen lone-pair orbital (nO), BE =10.90 eV. Within the COOH group of amino acids, one nonbonding orbital (nO) is localized on the oxygen atom of the hydroxyl group and another bonding orbital (πCO) is localized on the oxygen atom of the carbonyl group. Hence the peak observed in the VB spectrum of methionine at 12.0 eV was assigned to πCO orbital on the carboxylic group. These three peaks correspond very well to peaks in the Gly spectrum, where they are at about 0.2 eV higher energy [32]. They are typical of all amino acids, since they all contain an amino group (with nitrogen lone pair, nN), and a carboxylic acid group with oxygen derived molecular orbitals (nO, πCO).

The broad features in the valence band spectrum of Met between 12.5-20.3 eV (see Figure 3) can be assigned to ionization of several σ-type orbitals formed by the 2p orbitals of

C, N, and O atoms [34]. They show a different intensity ratio compared with the Gly spectrum, as well as an additional feature at 19.2 eV, where the Gly spectrum showed a deep minimum. This peak at 19.2 eV has been assigned to S 3s emission, possibly overlapping other C 2s derived states. In dimethyl sulfide the bonding environment of the sulfur atom is similar, and Kishimoto et al.. [35] found a peak at 19.02 eV assigned to S 3s and C 2s emission. Since methionine is a larger molecule than dimethyl sulfide, a small shift may be expected. The three peaks centered at 20.9, 22.5, and 23.85 eV have mostly C 2s character, and in contrast the Gly spectrum shows only two peaks at 20.2 and 23.3 eV in this BE region. The peak in the VB spectrum of methionine centered at ~28.2 eV is associated with N 2s ionization. The very broad maxima appearing at ~32.4 eV and ~34.3 eV (see Figure 3) correspond to the ionization of O 2s orbitals and again the BEs of these orbitals are similar to those observed in the Gly spectrum [33, 35].

Table 1. Valence band energies for methionine.

Binding energy (eV), experimental [36]	8.65, nS	9.8, nN	10.9, nO	12.0, πCO
Binding energy (eV), experimental [32]	8.65	9.8	10.9	12.1
Binding energy (eV), theoretical [33], conformers CF1, CF2(1,2)	8.09 CF1 8.60 CF2(1) 8.58 CF2(2)	9.63 CF1 9.72 CF2(1) 9.77 CF2(2)	10.61 CF1 10.81 CF2(1) 10.89 CF2(2) 10.91 CF1 11.09 CF2(1) 11.08 CF2(2)	11.80 CF1 11.29 CF2(1) 11.02 CF2(2) 11.72 CF1 12.05 CF2(1) 12.18 CF2(2)

Photoionization Mass Spectra

The mass spectra of Met as a function of photon energy are shown in Figure 5. There are two points to be noted about the lowest energy, 8.43 eV. This is lower than the first ionization potential so one would expect no ionization occurs. However the binding energy is not the minimum energy for ionization (known the "adiabatic IP"), but the average energy of the band, known as the "vertical IP." Thus ionization is possible, corresponding to the low energy tail of the peak. Secondly the light source used may emit light at a higher energy, 9.57 eV, and this may also affect the spectrum. At 8.43 eV, mostly the HOMO is ionized [32], and the dominant peak is the parent ion, m/z = 149. (Sulfur consists of about 96% of the isotope ^{32}S and 4% of the isotope ^{34}S, so sulfur containing peaks or fragments are accompanied by small peaks at two mass units higher with 4% of the main peak intensity.) As pointed out by Biemann et al.. [37] and later by Junk and Svec [38], the methylthio group can stabilise positive charge, thereby reducing fragmentation. Since the HOMO has mostly sulfur lone pair character, this explains why the parent ion is stronger than in other amino acids such as glycine, proline, etc. [24, 39-41]. The next most intense ions are m/z = 101 (loss of CH_3SH), 104 (loss of COOH), 132 (loss of NH_3) and 133 (loss of NH_2), whose loss mechanisms were discussed by Junk and Svec [38]. The first two processes correspond respectively to breaking of bond D in Fig. 5, with hydrogen transfer, and breaking of bond A in the same figure. The presence of these fragment ions at a photon energy where primarily the sulfur lone pair orbital is ionized underlines the fact that a molecular orbital is not entirely localized but may have significant weight at other sites in the molecule. Loss of COOH is believed to be facile due to resonance stabilization of the charge on the nitrogen atom, but is less likely to occur for charge localized on S, with three intervening atoms. The ion at m/z = 131 (loss of H_2O) appears, and it fragments further to produce an ion at m/z =83 ($C_4H_5NO^+$ loss of CH_3SH neutral fragment) [38] and m/z =84 ($C4H_6NO^+$, cleavage of the bond D).The peaks appearing at masses 115 and 116 could be due to the further fragmentation of m/z = 132 due to the loss of NH_2 and the m/z = 133 due to the loss of OH, respectively.

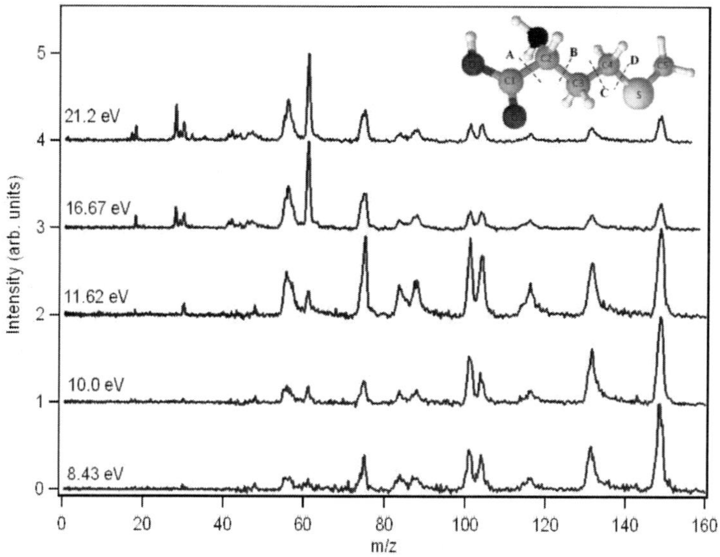

Figure 5. Mass spectra of methionine, as a function of photon energy.

Breaking of the bond B in Figure 5, without rearrangement gives two possible fragments of mass 74 ($NH_2CHCOOH$) and 75 ($CH_3SCH_2CH_2$). On the basis of isotopic distributions, Junk and Svec assigned m/z =75 to the second ion and m/z = 74 to the same ion with loss of hydrogen. There are weaker peaks at m/z = 86 and 87 which were observed at all photon energies due to the rupture of the C4–S bond and loss of NH_2 and NH_3 neutral fragments. The fragment at masses 87 ($C_3H_5NO_2^+$) and 88 ($C_3H_6NO_2^+$) are formed by simple cleavage of the C3–C4 bond (marked C in Fig. 5). The appearance of the most intense peak at higher incident photon energy (He I and Ne I) m/z =61 ($C_2H_5S^+$) in the mass spectrum of methionine is due to the direct cleavage of the C3–C4 bond, and it shows that most of the parent ions are probably produced after localization of the charge on the sulfur atom [38]. The ion $C_3H_6N^+$ at m/z = 56 is formed by loss of COOH and CH_3SH as proposed by Biemann et al.. [37]. Rupture of the bonds C1–C2 and C4–S (marked A and D in Fig. 5) gives rise to the fragment $C_3H_7N^+$ at m/z = 57, and with the loss two hydrogen atoms from this fragment, the ion $C_3H_5N^+$ (m/z=55) is formed. The peak at m/z=58 ($C_2H_4NO^+$, $C2H_2O_2^+$) may result from further fragmentation and loss of O or NH_2 from the amino acid fragment $C_2H_4NO_2^+$. The lower mass fragments at m/z =41, 42 and 44 are assigned to hydrocarbon ions. At 10.0 eV, both the HOMO and the HOMO-1 are ionized, but the mass spectra show only a small increase in fragmentation. The HOMO-1 has nitrogen lone pair character [32]. At 11.62, the three highest MOs are ionized, with the third highest orbital having oxygen lone pair character. There is a decrease in the yield of heavier ions, such as the parent ion and m/z =131–133 have been detected. This indicates that ionization of the oxygen lone pair orbital causes substantially more fragmentation than ionization of the nitrogen or sulfur lone pairs. At 16.67 eV, ionization of the 8 highest molecular orbitals occurs, and this is associated with strong fragmentation. A peak at m/z =61($C2H_5S^+$) becomes the strongest peak and peaks appear at m/z = 28–30. The spectrum at 21.2 eV is similar to the electron impact spectrum [42], and to Junk and Svec's spectrum [38].

By comparison with other amino acids, we can conclude that the photo-fragmentation process in methionine is different because the outer orbital has sulfur lone pair character. Ionization of a sulfur lone pair orbital evidently causes much less molecular fragmentation, while ionization of the nitrogen or oxygen lone pair leads to scission of the C1 to C2 bond, and generally more fragmentation.

Core Level Photoemission Spectra

In Figure 6, the carbon, nitrogen, and oxygen 1s as well as the sulfur 2p core level spectra of methionine are shown. These spectra are compared with those of Gly [36] and the core level binding energies are presented in Table 2.

Met contains four inequivalent carbon atoms so the C 1s spectrum was fitted with four Gaussian peaks (see Figure 6a), whose energies, widths, and intensities were free parameters. The first peak at 294.7 eV (295.2 eV for Gly) is assigned to the carboxylic carbon atom (C1) and the peak at 292.0 eV to the second carbon atom (C2) bonded to the amino group [36]. Methionine contains two more carbon atoms than glycine (see Figure 1), therefore the spectrum shows additional intensity in the lower binding energy range. The two additional peaks at 291.5 eV and 290.9 eV are due to the carbon atom bonded to two other carbon atoms (C3), and the carbon atoms bonded to the sulfur atom (C4, C5), respectively.

The N 1s spectra show a single asymmetric peak for methionine at 405.1 eV and for glycine at 405.4 eV and it is assigned to the nitrogen atom bonded to carbon atom C2. This shift is slightly larger than that of the N 2s level, underlining the different behavior of core and inner valence orbitals. It was found that this asymmetric shape of the N 1s spectrum of Gly, particularly the tail at higher binding energy, is due to contributions of the different molecular conformations of the molecules [8]. The conformers of amino acids in the gas phase are stabilized by hydrogen bonding, and the first type is characterized by an asymmetric NH2---O=C hydrogen bond (amino donor, oxygen acceptor), while the second type of conformer contains an OH---N hydrogen bond (oxygen donor, nitrogen acceptor). The feature at higher binding energy is less prominent in the N 1s spectrum of Met than for other amino acids we have measured, therefore we conclude that second type of conformer is not strongly populated [9, 36, 43]. Instead there may be minor effects due to conformers with and without hydrogen bonds between the amino (donor) and the sulfur (acceptor) atom.

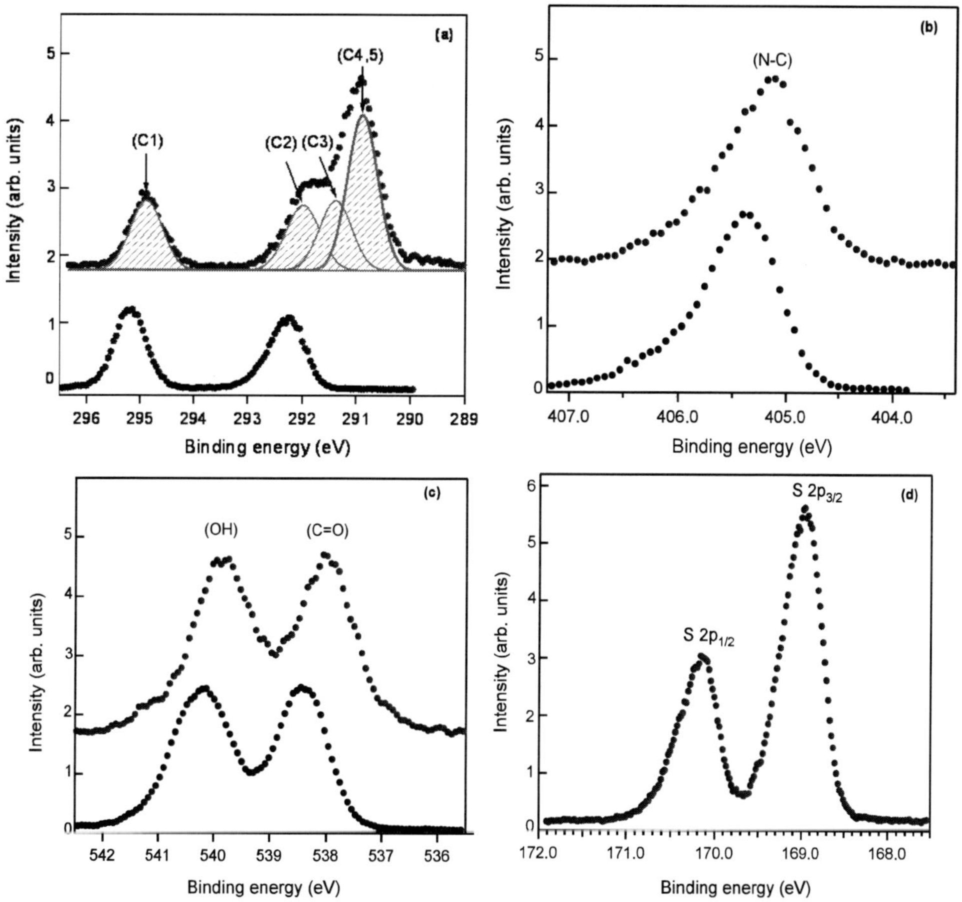

Figure 6. Core level spectra of methionine (top blue points) and glycine (lower black points): (a) Carbon 1s together with fitted curves and their assignment (see fig. 1), (b) Nitrogen 1s, (c) Oxygen 1s and (d) Sulfur 2p.

Table 2. C 1s, N 1s and O 1s binding energies for methionine.

Core level	Binding energies (eV)	Full Width at Half Maximum of the Gaussian fit function (eV), ±0.01 eV	Assignments	Intensity (relative to C(1)*=O or C(1)=O*)
C 1s	294.7	0.70	C(1)*=O	1
	292.0	0.86	N-C(2)*H_2	0.96
	291.5	0.86	C(3)*H_2	1
	290.9	0.87	S-C(4,5)*	2.08
N 1s	405.1	0.9 (asym)	H_2N*-C(2)	
O 1s	539.8	0.16 (asym)	C(1)-O*H	0.80
	538.0	0.16	C(1)=O*	1
S $2p_{3/2}$	169.0	0.56 (asym)	S*-C(4,5)	
S $2p_{1/2}$	171.2	0.61		

[a] asym. in the width column indicates a markedly asymmetric peak.

The two peaks in the oxygen 1s photoemission spectrum of methionine (see Figure 6(c)) are located at similar BE to those of Gly, and are assigned accordingly [36, 44]. The carbonyl oxygen atom gives rise to the peak at 538.0 eV while the hydroxy oxygen atom is the origin of the peak at 539.8 eV. The binding energy of S $2p_{3/2}$ in methionine, 169.0 eV, is equal within the precision of the measurement to the value found in dimethyl sulfide, 169.05±0.04 [45], as expected because the chemical environment of the sulfur atom is very similar.

C, N, and O K-Edge NEXAFS Spectra

Figure 7 presents the C, N and O 1s photoabsorption spectra of methionine, while the experimental energies of the resonances and proposed assignments are summarized in Table 3. The "term value" given in the Table is the difference between the resonance energy and the binding energy of the atom excited. In a simple model it can be regarded as the energy of the unoccupied orbital below the vacuum level. The NEXAFS spectra of Met are very similar to those of glycine, for which the gas phase NEXAFS spectra at all three edges have been calculated in detail [46].

Table 3. Energies of X-ray Absorption Peaks at the C, N and O 1s edges of methionine.

Edge	Energy (eV)	Assignment	Term value (eV)	Literature values [10]
C 1s	287.1	C4,5→σ*CHS	3.8	287.2 (3.5)
	287.6	C4,5→σ*CHS	3.3	287.9 (3.5)
	288.5	C1→π*	6.2	288.6 (6.4)
	289.2	C3→σ*CC	2.8	289.3 (2.9)
	290.7	C2→σ*CNH+Rydberg	-	
N 1s	401.2	σ*NH	3.9	
	402.4	π*NC	2.7	
	403.5	σ*NC	1.7	
	404.8	Rydberg states, shape resonances	0.3	
O 1s	532.2	O(CO)→π*	5.8	
	535.4	O(OH)→σ*(OH) O(OH)→σ*(OC) O(OH) 1s→3p	4.4	

The C 1s spectrum (see Fig. 7(a)) of methionine shows a prominent peak at 288.5 eV, and it has been assigned to excitations of the carboxylic carbon 1s to the empty π*(C=O) orbital. The two shoulders below it have energies of 287.1 and 287.6 eV and there is also some intensity at about 289.2 eV and a broad feature at about 290.7 eV. The methionine C K edge spectrum can be assigned by comparison with the solid state spectrum [10], because we expect the electronic structure of the carbon skeleton to be substantially preserved in the solid. The gas phase and solid state spectra agree rather well as the first four peaks are reproduced, and three out of the four peaks have the same energy if the energy scale is shifted rigidly by 0.1 eV. There is then a residual energy difference of 0.2 eV for the second peak. These peaks are therefore assigned following Kaznacheyev et al.. [10], and the lowest energy resonances have sulfur antibonding character. If the upper states are identified as unoccupied orbitals, then to decide which is the LUMO we need to examine the term values, as these take account of the variable binding energy of the core levels. The largest term value, 6.2 eV, is for the C1→π* transition, so this suggests that the LUMO is in fact the antibonding C1=O π orbital. This is supported by the O K edge NEXAFS spectrum, where the term value for corresponding transition is 5.8 eV. The small difference in energy reflects the fact that different cores have been excited.

Lastly, for the C edge, we observe a fifth peak at 290.7 eV which we assign to transitions to σ*CN and Rydberg states, by analogy with glycine.

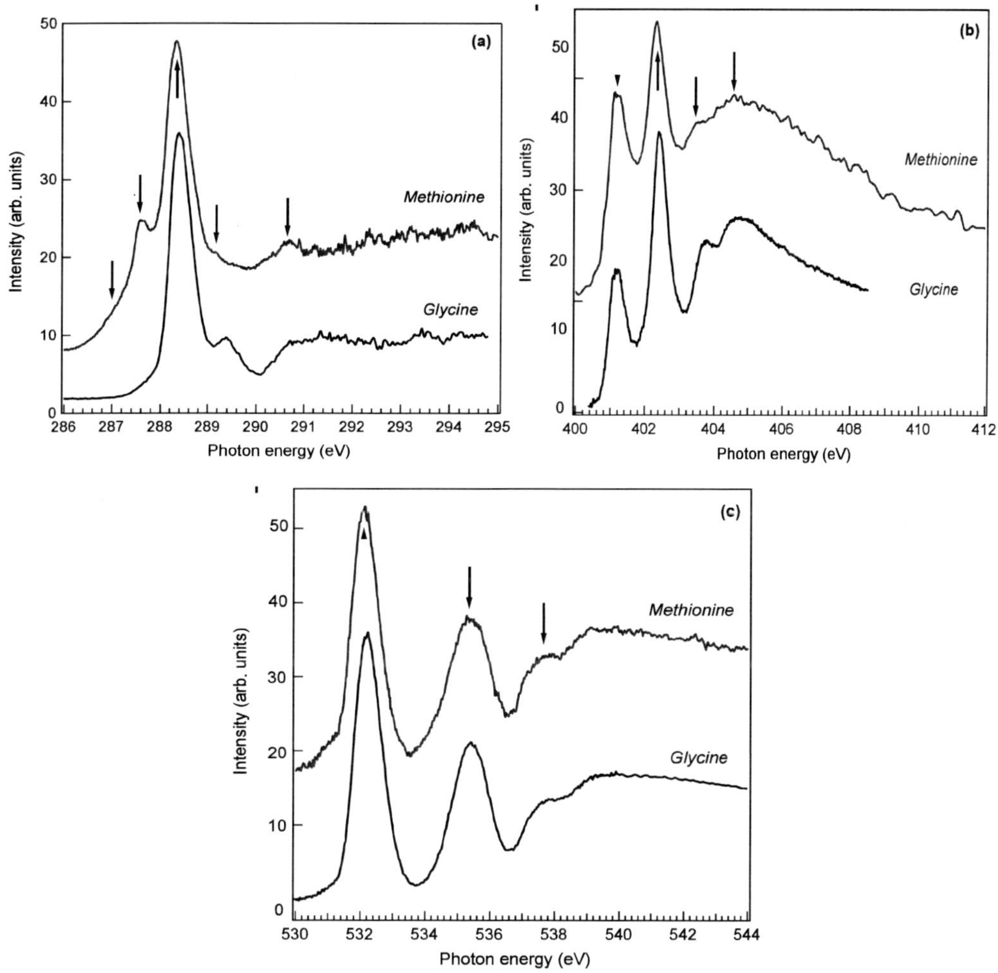

Figure 7. NEXAFS spectra of methionine: (a) C K-edge, (b) N K-edge and (c) O K-edge. Bars mark the positions of resonances.

The N K-Edge NEXAFS spectrum of methionine shows peaks at 401.2, 402.4 and about 405 eV (see Figure 7(b)). The first two have widths of 0.5 eV, while the third peak is very broad. Since the local environment of the nitrogen atom is similar to glycine, it is straightforward to assign the three peaks to the same transitions, as the term values are rather similar. Thus the maxima from the N 1s spectrum are due to the σ*NH, π*NC, and σ*NC transitions (see Table 3). The fourth peak is assigned to Rydberg states and mainly shape resonances. Note that there are no transitions to empty states of sulfur character: this is because the nitrogen atom is far from sulfur and the empty states on sulfur are localized. This is not always so: molecular orbitals may extend over several atoms, but that is not so here.

The oxygen K-edge spectrum of Met is shown in Figure 7(c) and it resembles Gly [46] and we make the same assignments as for glycine and formic acid [47]. The first narrower (~1.0 eV wide) peak at about 532.2 eV was assigned to oxo O(C=O) 1s transitions to π^* orbitals. The second broader (~1.5 eV) peak centered at about 535.4 eV is assigned to transitions from hydroxyl O(OH) 1s to the same vacant orbital and the mixed 3s/σ^* orbitals. The energy of the third peak matches the energy of the resonances of glycine and formic acid and is assigned to the O(OH) 1s$\rightarrow$3p transition [47, 48]. The theory predicts well the energies of the resonances, and the assignments are consistent with that given for formic acid [47, 48]. There are however some subtle differences. The O(OH) 1s$\rightarrow$3p transition at 538.4 eV in formic acid appears to shift to lower energies of about 537.5 eV and is weaker for methionine and the previously studied amino acids [8, 9, 46].

Conclusion

Methionine has been investigated by electron spectroscopy, x-ray absorption and photoionization mass spectroscopy. The frontier orbitals, which are those most involved in chemical reactions are different from glycine and many other amino acids because the HOMO is largely of S lone pair character. Low energy ionization of methionine then leads to less fragmentation of the molecule because the positive charge can be localized there and the ion is stable. We may therefore expect higher resistance to radiation damage for this component of proteins than for other amino acid residues, because the radiation resistance is a property of the side chain, rather than the groups which form the peptide bond, and change their chemical nature on protein formation. The other frontier orbitals of methionine follow the sequence of orbitals found in the amino acid used for comparison, glycine.

The spectra of the C, N, O and S core levels have been measured and assigned to the specific atoms in the molecule. This data was combined with the NEXAFS data to extract the term values, and draw conclusions about the unoccupied density of states. The lowest energy excitations at the carbon edge are to empty orbitals of sulfur character, but the LUMO is the π^* antibonding orbital of the carbonyl group in the carboxylic acid moiety.

Acknowledgment

V. C. would like to acknowledge the contribution of the "Ministero della Istruzione della Universitá e della Ricerca" of the Republic of Italy.

References

[1] R. B. Gerber, G. M. Chaban, S. K. Gregurick, B. Brauer, *Biopolymers*, 68, 370 (2003).

[2] M. S. de Vries, P. Hobza, *Annu. Rev. Phys. Chem.* 58, 585 (2007).

[3] C. Lomenech, G. Bery, D. Costa, L. Stievano, J.F. Lambert, *Chem. Phys. Chem.* 6, 1061 (2005).

[4] S.G. Stepanian, I.D. Reva, E.D. Radchenko, M.T.S. Rosado, M.L.T.S. Duarte, R. Fausto, L. Adamowicz, *J. Phys. Chem. A* 102, 1041 (1998).

[5] S.G. Stepanian, I.D. Reva, E.D. Radchenko, L. Adamowicz, *J. Phys. Chem. A* 105, 10664 (2001).

[6] Y. Zubavichus, A. Schaporenko, M. Grunze, M. Zharnikov, *J. Phys. Chem. A* 109, 6998 (2005).

[7] Y. Zubavichus, M. Zharnikov, A. Shaporenko, O. Fuchs, L. Weinhardt, C. Heske, E. Umbach, J. D. Denlinger, M. Grunze, *J. Phys. Chem. A* 108, 4557 (2004).

[8] V. Feyer, O. Plekan, R. Richter, M. Coreno, K. C. Prince, V. Carravetta, *J. Phys. Chem. A* 113, 10726 (2009).

[9] V. Feyer, O. Plekan, R. Richter, M. Coreno, K. C. Prince, V. Carravetta, *J. Phys. Chem. A* 112, 7806 (2008).

[10] K. Kaznacheyev, A. Osanna, C. Jacobsen, O. Plashkevych, O. Vahtras, H. Ågren, V. Carravetta, A. P. Hitchcock, *J. Phys. Chem. A* 106, 3153 (2002).

[11] M. Tanaka, K. Nakagawa, T. Koketsu, A. Agui, A. Yokoya, *J. Sync. Radiat.* 8, 1009 (2001).

[12] F. Kaneko, M. Tanaka, S. Narita, T. Kitada, T. Matsui, K. Nakagawa, A. Agui, K. Fujii, A. Yokoya, *J. Electron Spectrosc. Relat. Phenom.* 144-147, 291 (2005).

[13] J. Boese, A. Osanna, C. Jacobsen, J. Kirz, *J. Electron Spectrosc. Relat. Phenom.* 85, 9 (1997).

[14] O. Plachkevytch, V. Carravetta, O. Vahtras, and H. Agren, *Chem.Phys.* 232, 49 (1998).

[15] V. Carravetta, O. Plashkevych, and H. Agren, *J. Chem. Phys.* 109, 1456 (1998).

[16] L. Yang, O. Plashkevych, O. Vahtras, V. Carravetta, and H. Agren, *J. Synchrotron Radiat.* 6, 708 (1999).

[17] T. L. Brown, H. E. Jr. Lemay, B. E. Bursten, *Chemistry the Central Science*, 6th ed.; Prentice-Hall: Englewood Cliffs, NJ, 1994; Chapter 26, p 982.

[18] K. J. Barnham, M. I. Djuran, P. D. S. Murdoch, J. D. Ranford, P. J. Sadler, *J. Chem. Soc., Dalton Trans.* 3721 (1995).

[19] R. Linder, K. Seefeld, A. Vavra, K. Kleinermanns, *Chem. Phys. Lett.* 453, 1 (2008).

[20] A. Mijovilovich, L. G. M. Pettersson, S. Mangold, M. Janousch, J. Susini, M. Salome, F. M. F. de Groot, and B. M. Weckhuysen, *J. Phys. Chem. A* 113, 2750 (2009).

[21] S. Desaphy, C. Malosse, G. Bouchoux, *J. Mass Spectrom.* 43, 116 (2008).

[22] K. C. Prince, R. R. Blyth, R. Delaunay, M. Zitnik, J. Krempasky, J. Slezak, R. Camilloni, L. Avaldi, M. Coreno, G. Stefani, C. Furlani, M. de Simone, S. Stranges, *J. Synchrotron Radiat.* 5, 565 (1998).

[23] O. Plekan, V. Feyer, R. Richter, M. Coreno, M. de Simone, K. C. Prince, A. B. Trofimov, E. V. Gromov, I. L. Zaytseva, J. Schirmer, *Chem. Phys.* 347, 360 (2008).

[24] O. Plekan, V. Feyer, R. Richter, M. Coreno, M. de Simone, K. C. Prince, *Chem. Phys.* 334, 53 (2007).

[25] J. Jen and T. D. Thomas, *J. Electron Spectrosc. Relat. Phenom.* 4, 43 (1974).

[26] L. Petterson, J. Nordgren, L. Selander, C. Nordling, K. Siegbahn, *J. Electron Spectroscopy Relat. Phenom.* 14, 29 (1982).

[27] V. Myrseth, J. D. Bozek, E. Kukk, L. J. Sæthre, T. D. Thomas, *J. Electron Spectrosc. Relat. Phenom.* 122, 57 (2002).
[28] T. D. Thomas and R. W. Shaw, *J. Electron Spectrosc. Relat. Phenom.* 5, 1081 (1974).
[29] M. Tronc, G. C. King, F. H. Read, *J. Phys. B: At. Mol. Phys.* 12, 137 (1979); *J. Phys. B: At. Mol. Phys.* 13, 999 (1980).
[30] R. N. S. Sodhi, C. E. Brion, *J. Electron Spectrosc. Relat. Phenom.* 34, 363 (1984).
[31] G. R. Wight, C. E. Brion, *J. Electron Spectrosc. Relat. Phenom.* 3, 191 (1974).
[32] P. H. Cannington, N. S. Ham, *J. Electron Spectrosc. Relat. Phenom.* 32, 139 (1983).
[33] D. Dehareng, G. Dive, *Int. J. Mol. Sci.* 5, 301 (2004).
[34] C. T. Falzon, F. Wang, W. Pang, *J. Phys. Chem. B* 110, 9713 (2006).
[35] N. Kishimoto, R. Yokoi, H. Yamakado, K. Ohno, *J. Phys. Chem. A* 101, 3284 (1997).
[36] O. Plekan, V. Feyer, R. Richter, M. Coreno, M. de Simone, K. C. Prince, V. Carravetta, *J. Phys. Chem. A* 111, 10998 (2007).
[37] K. Biemann, J. Seibl, F. Gapp, *J. Am. Chem. Soc.* 83, 3795 (1961).
[38] G. Junk and H. Svec, *J. Am. Chem. Soc.* 85, 839 (1963).
[39] F. Lago, L. H. Coutinho, R. R. T. Marinho, A. Naves de Brito, and G. G. B. de Souza, *Chem. Phys.* 307, 9 (2004).
[40] L. H. Coutinho, M. G. P. Homem, R. L. Cavasso-Filho, R. R. T. Marinho, A. F. Lago, G. G. B. de Souza, A. Naves de Brito, *Braz. J. Phys.* 35, 940 (2005).
[41] H.-W. Jochims, M. Schwell, J. L. Chotin, M. Clemino, F. Dulieu, H. Baumgartel, and S. Leach, *Chem. Phys.* 298, 279 (2004).
[42] NIST, Chemistry WebBook. http://webbook.nist.gov/chemistry.
[43] W. Zhang, V. Carravetta, O. Plekan, V. Feyer, R. Richter, M. Coreno, and K. C. Prince, *J. Chem. Phys.* 131, 035103 (2009).
[44] R. Slaughter, M. S. Banna, *J. Phys. Chem.* 92, 2165 (1988).
[45] W. L. Jolly, K. S. Bomben, C. J. Eyermann, *At. Data and Nuclear Data Tables* 31, 433, (1984).
[46] O. Plekan, V. Feyer, R. Richter, M. Coreno, M. de Simone, K. C. Prince, V. Carravetta, *J. Electron Spectrosc. Relat. Phenom.* 155, 47 (2007).
[47] K. C. Prince, R. Richter, M. de Simone, M. Coreno, *J. Phys. Chem. A* 107, 1955 (2003).
[48] D. Di Tommaso, P. Decleva, *J. Chem. Phys.* 123, 064311 (2005).

In: Methionine
Editor: Alexander Snegursky

ISBN: 978-1-62618-246-2

Chapter VI

Photoluminescence of Methionine under Laser Excitation

Maria I. Myhovich, Yuriy O. Shpenik and Volodymyr A. Kelman [*]
Institute of Electron Physics, National Academy of Sciences of Ukraine, Ukraine

Abstract

Photoluminescence of methionine and absorption spectra of its aqueous solution were measured using laser excitation at three different wavelengths (253, 337, and 380 nm) as well as a UV deuterium lamp. With the excitation laser wavelength increase, the photoluminescence short-wave edge and the main maxima appear to shift towards the red. In a number of cases Raman scattering by water was observed along with the luminescence spectra. Intensity ratio of the Raman and the photoluminescence bands can serve a natural scale to compare the photoluminescence efficiency measured in different experiments and the humidity of the amino acid powder. The long-wave edge in the absorption spectra gives a clear evidence that the properties of the methionine molecules in aqueous solutions are close to those in the solid state. Some energy parameters of methionine were obtained.

Introduction

Amino acids are organic compounds composed of molecules which include an alkaline amino group (-NH_2) and an acidic carbonyl group (-COOH). Often they can exist in the form of so-called zwitter-ions. They are the main constituents of proteins and are involved in all vital processes, along with nucleic acids, carbohydrates, and lipids.

[*] Corresponding author: E-mail address: vkel@rambler.ru.

Methionine amino acid $C_5H_{11}NO_2S$ (molar mass 149.21 g/M) with side chain R polarity refers to non-polar amino acids. With a lack of carbohydrates in a body, methionine could be converted into glucose or glycogen through a chain of reactions. Methionine is one of the essential amino acids provided to the human body from the food. The chemical structure of methionine is shown in Figure 1.

The electronic structure of molecules can be investigated by measuring absorption and luminescence spectra of the corresponding molecules. Luminescence study of substances enables one not only to discuss the state of molecules in the system, but also to trace the energy transfer between the molecules. The measured positions of maxima in the spectral dependences of photoluminescence (PL) intensity enable one to judge upon the electronic excitation levels, in which the molecule can store the absorbed energy. Besides, luminescence spectra of complex systems enable one to trace the migration of an absorbed quantum of energy between different molecules or groups of other molecules, as well as the energy accumulated at metastable states.

To study the energy structure of methionine we have carried out an experimental investigation of its PL spectra excited by laser sources of different wavelengths.

It should be mentioned that we are not aware of other direct studies of methionine PL. Methionine is also absent in a well-known PHOTOCHEMCAD software database. Apparently there are only a few works indirectly related to the luminescence of methionine In particular, water dispersible ZnS:Mn nanocrystals were synthesized by capping the surface of the nanocrystals with four kinds of amino acid ligands: arginine, cysteine, histidine, and methionine [1]. The optical properties of the amino acid capped ZnS:Mn colloidal nanocrystals were measured using the UV/Vis and photoluminescence spectroscopies of aqueous solutions. The PL spectra of the solutions show broad emission peaks around 575 nm (orange emission) with PL efficiencies in the range of 4.4 to 7.1%. The measured particle sizes for the amino acid capped ZnS:Mn nanocrystals were in the range of 5.3 to 11.7 nm.

A new analytical method for determining cysteine and methionine molecules was presented in [2]. A linear relationship between the changes in the electric conductivity oscillation period of the H_2O_2–KSCN–$CuSO_4$ chemical system and the concentration of cysteine and methionine admixture was observed. Monitoring of the conductivity oscillation frequency of the system was performed using a simple analytical method, namely, a conductometry technique. Linear dynamic range was obtained in the methionine concentration range of 2.6×10^{-4} M – 1.1×10^{-3} M.

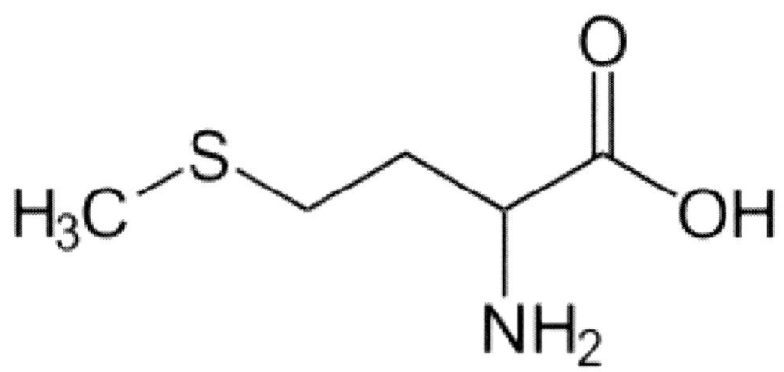

Figure 1. Chemical structure of a methionine molecule.

Self-assembly of L-methionine amino acid on Cu(111) was investigated under ultrahigh vacuum (UHV) conditions [3] by scanning tunneling microscopy (STM), helium atom scattering (HAS), and X-ray photoelectron spectroscopy (XPS). The system is strongly influenced by the substrate reactivity and the deposition temperature. The STM and HAS structural analysis yields that, for temperatures below 273 K, the biomolecules assemble in strings oriented at an angle of -10° with respect to the (110) axis of the substrate. For temperatures above 283 K, a regular and ordered one-dimensional (1D) phase arises at an angle of +10° with respect to the same direction. High resolution STM data of this ordered 1D arrangement give the evidence for molecular dimerization and dimer alignment into ordered chains which are commensurate with the Cu(111) atomic lattice. The XPS measurements reveal that the high temperature ordered phase consists of an exclusively anionic ensemble with a deprotonated carboxylic group and a neutral amino group, while the low temperature phase is heterogeneously composed of both zwitterionic and anionic species, depending on whether the molecules are immobilized in clusters of dimers on the free terraces or at low-coordinated adsorption sites of the substrate step-edges. These combined results give the evidence for a structural transformation of the supramolecular assembly which is triggered by a thermally activated process involving the underlying Cu(111) substrate and carries the intrinsic chiral signature of the adsorbed molecular units.

Experimental Apparatus and Method

We used selective excitation of luminescence of the methionine targets in the forms of a crystalline powder and an aqueous solution. These targets were illuminated using different laser sources such as a molecular nitrogen laser (λ = 337 nm) or a tunable Ti:Sapphire laser CF 131A (the second harmonic λ = 380 nm, the third harmonic λ = 253 nm). The laser pulse length was 5–20 ns. The pulse repetition frequency was 10 Hz for 253 and 380 nm, 100 Hz for 337 nm. The pulse energies did not exceed 0.5 mJ (253 nm), 0.1 mJ (337 nm), and 2 mJ (380 nm). It was supposed that the obtained complementary data using different laser sources would give an important information about the substances under study.

Liquid and solid methionine samples were illuminated directly by an unfocused laser beam with a diameter of 2 mm into a quartz cuvette containing the methionine target (Figure 2). The cuvette width was 2 mm for the powder sample and 12 mm for the solution, In all cases the images of the luminescence area with a scale 1:1 were projected on the entrance slit of an MS 7504i monochromator (grating grid 150 gratings/mm, reciprocal dispersion 8.78 nm/mm) using a quartz lens with f=75 mm. In front of the monochromator entrance slit the following filters were installed: BS-3 for the 253 nm excitation, ZHS-10 for the 337 nm, and ZHS-11 for the 380 nm. These filters reliably absorbed the scattered light of the laser excitation lines and transmitted the luminescence emission of the methionine samples. The photoluminescence spectra were recorded using an HS 101H CCD-camera and a PC. The entrance slit width in most cases was 0.5 mm and the signal acquisition time was 30 s. All the measurements were carried out at room temperature. During the measurement period we did not observe any essential changes in the photoluminescence spectra and, therefore, it should be believed that destruction of the methionine probes and, accordingly, storage of the degradation products during the measurement time were negligible.

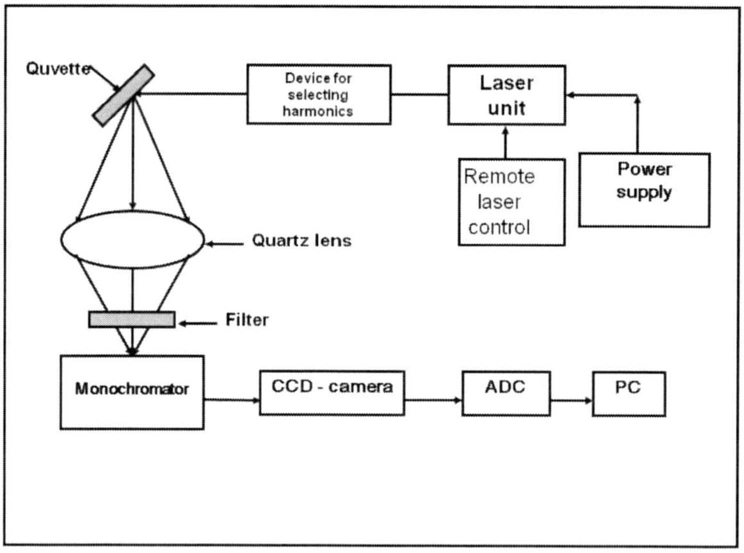

Figure 2. Experimental setup.

The concentration of the methionine aqueous solution was 0.067 M/l. The solution was prepared in order the absorption not to be too large at the depth of about 4 cm of the cuvette used, and the luminescence intensity to be the same along the luminescent path. This pathwas directly adjacent to the output window of a quartz cell to prevent re-emission of the luminescence and distortion of the spectral dependence of the photoluminescence intensity.

A DDC-30 series deuterium lamp was used to measure the absorption spectra of the methionine aqueous solution in the near UV spectral region. Then these absorption data were used to calculate the molecular extinction coefficient for methionine.

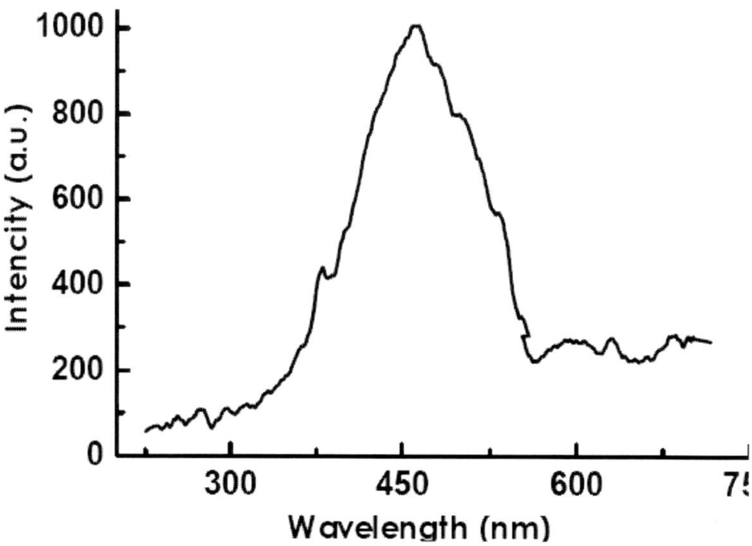

Figure 3. PL spectrum of the methionine aqueous solution excited by the third harmonic (λ=253 nm) of a Ti:sapphire tunable laser.

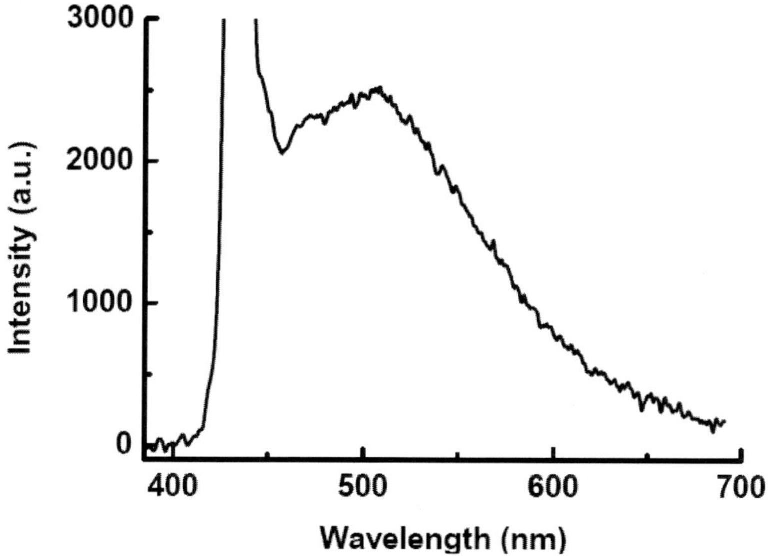

Figure 4. PL spectrum of methionine aqueous solution excited by the second harmonic (λ=380 nm) of a Ti:sapphire tunable laser.

Results and Discussion

We have measured only four photoluminescence spectra of the methionine targets of the six initially planned. The reason for this is that the methionine luminescence intensity was too low as compared to that of aromatic amino acids. The luminescence spectra of the liquid target are presented in Figures 3 and 4. The photoluminescence intensities are shown in arbitrary units. The relative sensitivity of the spectral arrangement has also been taken into account. As a rule, while analyzing the photoluminescence spectra we took into account the luminescence onset at the short-wave edge and the position of the main maximum.

Here one can see that with the pumping laser wavelength increase, the luminescence onset from the short wavelength edge and the main maximum are both shifted to the red. A distinct maximum peaking at 434 nm (intensity 13100 a.u.) in Figure 4, observed at the excitation with the 380 nm laser wavelength, corresponds to the Raman scattering by water molecules. As it is known, the frequencies of symmetric stretch ν_1 and asymmetric stretch ν_2 vibration bands in a liquid water are about 3280 cm^{-1} and 3490 cm^{-1}, respectively [4]. For the pumping laser wavelength of 380 nm such Raman shift value corresponds to the maximum observed at 434 nm. Sometimes we were able to observe the discussed peak even for the powder targets with essential humidity (see, e. g., Figure 6). Thus, comparing the intensities of the Raman peak and the photoluminescence, we can have a natural scale to compare the photoluminescence efficiency measured in different experiments.

Luminescence spectra of the powder target are presented in Figures 5 and 6. Here one can see that the general features in the spectral dependences are similar to those for the liquid. Comparing the spectra in Figures 4 and 6, both obtained with the 380 nm laser excitation, one can note that the short wave edge and the main maximum of photoluminescence for the powder sample are slightly shifted to the red region. A Raman peak could also be observed here at 425 nm.

All the above main features of the measured spectra are summarized in Table 1 that also shows the excitation laser wavelengths and the spectral positions of the main intensity maxima, and relevant energy.

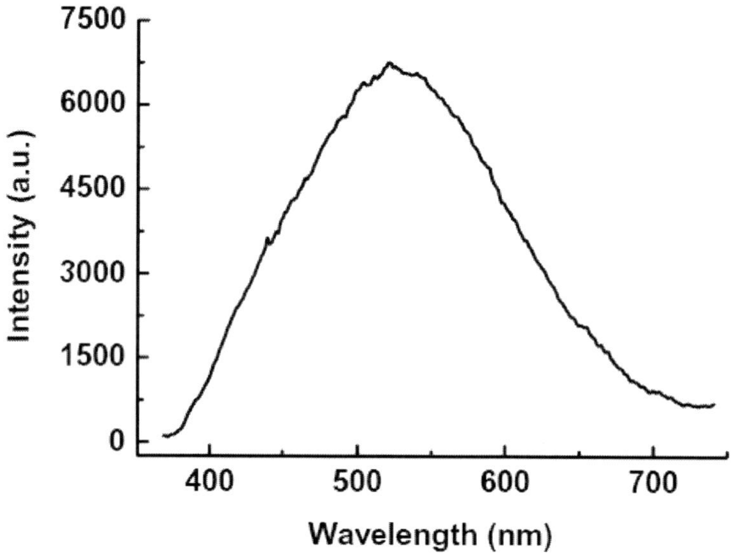

Figure 5. PL spectrum of the methionine powder excited by a nitrogen laser (λ=337 nm).

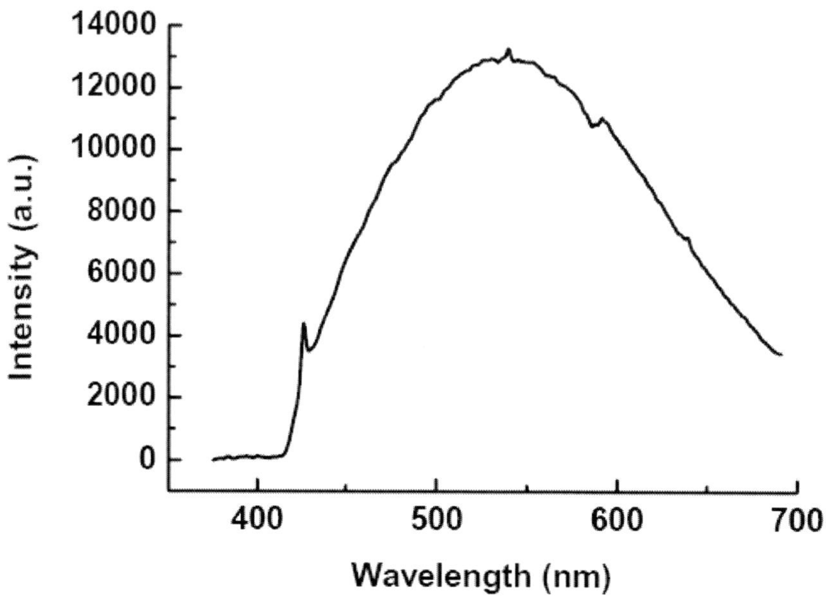

Figure 6. PL spectrum of the methionine powder excited by the second harmonic(λ=380 nm) of a Ti: Sapphire tunable laser.

Table 1. Spectral positions of the main luminescence maxima of methionine

Target	Excitation laser		Photoluminescence		
	λ_{exc}, nm	E_{exc}, eV	λ_{ph}^{max}, nm	I, a.u.	E_{ph}^{max}, eV
Aqueous solution	253	4.90	463	1007	2,67
	380	3.26	502	1914	2,47
Powder	337	3.68	521	6755	2,38
	380	3.26	539	2500	2,30

Finally, the energy dependence of the methionine molar extinction coefficient in the aqueous solution is shown in Figure. 7. This dependence was obtained by measuring the absorption spectra with subsequent calculation using the expression $\xi = -\frac{1}{cl}\ln\tau$. Here $\tau = \frac{I}{I_0}$ is the transmittance of the solution; I and I_0 are the intensities of the light passing through the cuvette with the aqueous solution and the incident light, respectively, c is the concentration of the solution (M/l), l is the thickness of the cuvette in cm.

It should be noted that such dependence is very similar to those characteristic for solids. When one extrapolates the linear part of the graph in figure 7 to the abscissa, then the energy gap of 4.85 eV and the respective long wave edge of absorption band of 255 nm can be estimated. So we can argue about a certain similarity of methionine properties in the form of powder and in the aqueous solution.

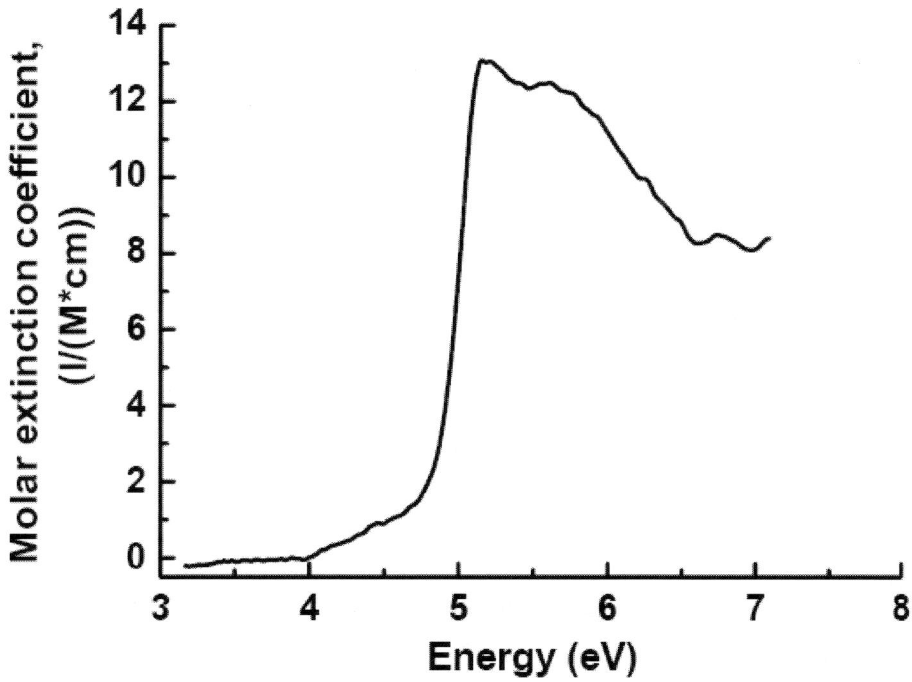

Figure 7. Energy dependence of the molar extinction coefficient for the methionine aqueous solution.

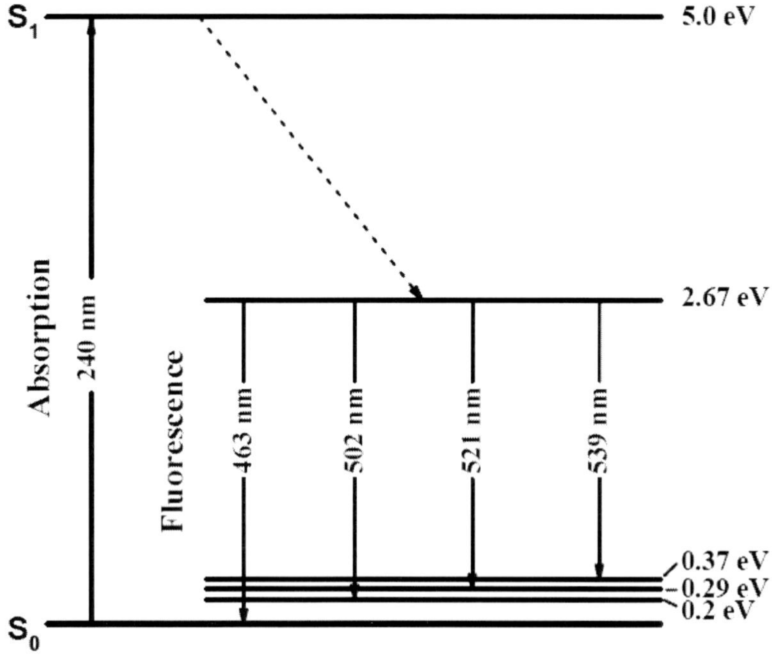

Figure 8. Scheme of methionine electron states and transitions.

Meanwhile, the absorption edge is somewhat smeared. This indicates that the aqueous solution of methionine is a disordered matter and there are localized electron states in the gap. The density and especially the location of the above states determine the character and the shape of the absorption edge. This problem should be studied in more detail in the future.

However, there are still some questions related to the photoluminescence spectra obtained. First, how can photons with energies corresponding to the wavelengths of 337 nm and, moreover, 380 nm excite the methionine molecules? It can be seen from figure 7, absorption at these wavelengths is rather small being far away from the main absorption band near 240 nm. Nevertheless, the luminescence is observed in this case. An assumption that in this case one observes a two-photon excitation process similarly to [5] does not seems probable. Indeed, looking at the dependence of the photoluminescence intensity on the exciting laser intensity, one can see rather a linear (not quadratic) character of the curve. Therefore, we believe that in our case the intensity of the luminescence does not follow the consistent pattern [6] established for the aromatic amino acids, i.e. the photoluminescence intensity does not depend on the optical density of the target. We think that , we are able to observe the luminescence excited with the wavelengths of 337 and 380 nm, in particular, due to the large optical density of the targets. In addition, one cannot exclude that the excitation of methionine in this case results from excited vibrational states of the ground electron state.

Using conventional techniques and knowing the luminescence and absorption parameters (see Table 1), one can obtain the energy of the S_1 state (approximately 5 eV, corresponding to 240 nm at the highest absorption value). Then the energy (2.67 eV) of the state producing the radiative transitions at 463, 502, 521, and 539 nm and the energies of the lower states of these transitions (0, 0.2, 0.29, and 0.37 eV, respectively) could be estimated. The features of the energy structure of methionine are shown in Figure 8.

Conclusion

Present study was carried out being aimed at tracing the photoluminescence features of methionine amino acid in the forms of powder and aqueous solutionas well as its absorption, and subsequent determining certain energy parameters of the material. The luminescence excitation was performed using different laser sources (with the 253, 337 and 380 nm wavelengths).

It was found that the luminescence onset from the short wavelength edge and the main maximum both shift to the red when the pumping laser wavelength increases . At the excitation with 380 nm wavelength, the photoluminescence spectra of the liquid and powder samples are observed along with a maximum peaking at 434 nm corresponding to the Raman scattering by water molecules. Comparison of the Raman peak and the photoluminescence intensities provides a natural scale to compare the photoluminescence efficiency in different experiments. Based on these measurements the methionine energy diagram in the first approximation was obtained. We believe these results will be useful to identify methionine among other organic substances.

The authors are grateful to Prof. A.V. Snegursky for the helpful cooperation.

References

[1] Ju Ho Lee, Yong Ah Kim, Kimoon Kim, Young Duk Huh, June Won Hyun, Kim, H. S., Noh, S. J. & Cheong-Soo Hwang, (2007). *Bull. Korean Chem. Soc.,* 28, 1091.

[2] *Abdolraouf Samadi-Maybodi and Kambiz Abolfazli, Int. J. Electrochem. Sci.,* 4, 684 (2009).

[3] Schiffrin, A., Reichert, J., Pennec, Y., AuwËarter, W., Weber-Bargioni, A., Marschall, M., Dell'Angela, M., Cvetko, D., Bavdek, G., Cossaro, A., Morgante, A. & Barth, J. V. (2009). *J. Phys. Chem.* C, 113, 12101–12108.

[4] Eisenberg D. & Kauzmann, W. (1969). *The structure and properties of water,* Oxford University Press, London.

[5] Gorelik, V. S. & Kozulin, E. A. (1994). *Quantum Electron, 24* (5), 462.

[6] Albani, J. R. (2007). *J Fluoresc. 17*, 406.

In: Methionine
Editor: Alexander Snegursky
ISBN: 978-1-62618-246-2

Chapter VII

Methionine Molecule Electron-Impact-Induced Fragmentation: Mechanisms and Chemical Structure

Alexander V. Snegursky*[*1], *Jelena Tamuliene*[2], *Vasyl S. Vukstich*[1] and *Liudmila G. Romanova*[1]
[1]Institute of Electron Physics,
Ukrainian National Academy of Sciences, Uzhgorod, Ukraine
[2]Vilnius University, Institute of Theoretical Physics and Astronomy, Vilnius, Lithuania

Abstract

Present chapter deals with the complex experimental and theoretical studies dedicated to the electron–impact induced fragmentation of the methionine ($C_5H_{11}NO_2S$) molecule belonging to a wide class of amino acids. Experiments were carried out by a crossed-beam technique with the mass-spectrometric analysis of the collision products of the processes of dissociative ionization of the initial molecule using a modernized high-sensitivity magnetic mass analyzer. The principal details of the experimental procedure and theoretical approach applied are described. The main emphasis has been given to the production of a number of ionized fragments from the initial molecule under study. A vast theoretical analysis allowed the main fragmentation channels of the methionine molecule to be elucidated and the basic geometric characteristics of this molecule structure to be found.

Introduction

It is a well-established fact that interaction of ionizing radiation with live tissue causes irreversible effects at the genetic level [1]. The mechanisms of living cell material

[*] E-mail: snegursky.alex@gmail.com

degradation include not only the direct action of high-energy ionizing radiation but also indirect processes related to the chemical structure transformations under the influence of low-energy secondary electrons resulting from the incident ionizing radiation interacting with the target. The primary radiation itself induces approximately one third of the mutagenic transformations in the cell [2], whereas most of the above radiation-induced transformations are related to both the secondary electrons and free radicals produced. Secondary products of ionizing radiation in living tissues are capable of damaging the structural units of nucleic acids and proteins leading, in particular, to dissociative ionization and negative-ion production. Thus, one can state that low-energy electrons are of primary interest from the viewpoint of tracing the consequences of both malignant transformations in living cells under the influence of ionizing radiation and also a useful radiation therapy effect on tumors in human beings [3].

This specifies kind of interest in studying inelastic processes of slow-electron interactions with molecules of biological relevance. Such processes include, among others, electron-impact excitation, electron attachment (including dissociative one), direct and dissociative ionization resulting in reactions leading to the formation of positive ions and slow secondary electrons. The dissociative ionization is extremely important due to the production of the parent molecule fragments both in the ionized and neutral states. Among a wide variety of the above targets amino acids are peculiar not only due to the fact that they are the building blocks of the human body but also in view of the role they could have played in the genesis and development of the life on the Earth. Current research has shown the presence of amino acids in various space objects (say, meteorites), which, in principle, may prove the idea of 'life import' to our planet [4].

A very significant fact is that such molecules are involved in the processes of building cells and repairing damaged tissues as well as the processes of forming antibodies to struggle against foreign viruses and bacteria. Being a part of the hormonal and enzyme system, the molecules of amino acids are responsible for building the RNA and DNA nucleoproteins and for the oxygen transfer in the live tissue. Methionine ($C_5H_{11}NO_2S$, or $HO_2CCH(NH_2)CH_2CH_2SCH_3$), that belongs to the amino acid family and is one of the two sulfur-containing proteinogenic amino acids is known as an intermediate in the biosynthesis of carnitine, cysteine, lecithin, phosphatidylcholine, taurine, and other phospholipids. In addition, methionine is one of only two amino acids encoded by a single codon in the standard genetic code (together with tryptophan).

The studies of the basic mechanisms of the molecule structural changes caused by low-energy electron impact are far from being complete or consistent despite their undoubted significance. According to the National Institute of Standards (NIST) database [5], the available data on this problem are quite contradictory, while those on the parent molecule ionization thresholds and fragment appearance energies are very scarce. The main issue here is related to the fact that molecules of biological relevance, including aminoacids, undergo strong fragmentation due to both the effects of low-energy electron impact and thermal degradation in the process of molecular flux formation during beam experiments. It is essential, therefore, to distinguish clearly the molecular fragments formed just due to dissociative ionization. The kinetic energy of the ionized fragments is of specific importance as well since it provides qualitative and quantitative estimates of the nature and extent of structural changes and transformations in the living cell. Of particular interest are also the data on the appearance energies of ionic fragments. We will turn to analyzing the data available on this problem when discussing the results, obtained by us in the course of our investigations (see below).

Experimental

In our study, we used a conventional magnetic mass-spectrometer MI-1201 as the mass-separator unit enabling the fragments of electron-molecule interaction to be identified. In its original design, this device is not capable of studying the energy dependences of the ion yield providing the possibility of measuring the mass spectra of the fragments at fixed ionization energies mainly due to the fact that the geometry of the electron beam in the conventional ion source strongly depends on the energy of electrons. This fact has led us to the idea of modifying the mass spectrometer by implementing the specially designed electron beam source (see below) what resulted in the possibility to study both ionization and dissociative ionization cross sections as the functions of incident electron energy and to determine the appearance energies of the fragments under study.

The experimental apparatus developed and applied in this chapter is based, as it was already mentioned, on the magnetic mass spectrometer capable of operating within the 1–720 a.m.u. mass range. High sensitivity (~10^{-16} A) and resolution ($\pm$0.25 a.m.u.) of the mass analyzer enabled the fragments of the target molecule to be reliably separated and detected even at low levels of ion currents reaching the ion detector. The primary molecular beam was formed by means of an effusion source with a resistive oven providing target molecule concentrations not less than 10^{10} molecule/cm^3. The operating temperature of the molecular beam source was varied up to 150°C and controlled by a thermocouple enabling the temperature dependences of the fragments yield to be determined. The temperature and pressure conditions of the molecular beam formation excluded the possibility of molecular cluster formation. A specially designed three-electrode electron gun provided an electron current of 30–50 μA over a wide energy range. This source enabled the energy dependences of the ionization and dissociative ionization cross sections to be measured in the incident electron energy range from the threshold up to 150 eV without the need of using a magnetic field (contrary to the conventional source). The ions, produced in the ion source and extracted by the electric field, entered the magnetic ion separator and were detected by means of an electrometer. The data acquisition and processing system were controlled by a PC. Special measures were taken to stabilize the mass analyzer transmission thus making the mass of the fragment under study to be reliably fixed. An electron energy scale was calibrated with respect to known ionization thresholds for argon atom and nitrogen molecule (see below) with the accuracy of $\pm$0.1 eV.

Modernization of the Ion Detection, Control and Mass-Spectra Sweep Channels

The modernized mass spectrometer has the following functional capabilities:

i. The increase in the ion-detection sensitivity by two orders of magnitude (to 10^{-16} A) and the decrease in the electrometer zero-level drift and fluctuation to 0.5 mV due to low-noise semiconductor solid-state electrometers used;
ii. The possibility of converting the voltage, picked off the electrometer into the pulse repetition rate with the conversion coefficient of 50 kHz/V for further digital signal processing by a computer;
iii. The stabilization of the mass analyzer magnetic-field induction and a precise pulsed mass-spectrum sweep with a minimum step of 10^{-4} a.m.u. owing to the implementation of a digital-sweep generator with a frequency-to-current converter and a pulse-generator temperature stabilizer;
iv. The application of a controlling computer with a specially designed applications program for the creation of an adjustment medium in which all the functional capabilities of the mass spectrometer are implemented through hierarchic menu elements;
v. The possibility of precise scale calibration within an arbitrary mass range according to two reference peaks (even with substantially different intensities);
vi. The possibility of multiple cyclic alternate centering of peaks with the subsequent determination of their relative intensities and errors;
vii. The possibility of saving the visualized mass spectra recorded in graphical and digital formats on a hard disk.

The above characteristics allowed one to promptly and comprehensively study the mass spectra of the fragments of the methionine molecules.

Modernization of the Ion Source and its Power Supply

Figure 1 shows the schematic diagram of the electron gun, control system of the electron-beam energy, and circuits for measuring the electron-accelerating voltage U_e and the beam current I_e. To modernize both the electron gun and the molecular beam source (see Fig. 1) one of the two gaseous ion sources [6] was reconstructed so as to preserve the possibility of recording mass spectra at any constant electron energy and to provide automatic measurements of the energy dependences of the ionization and fragmentation efficiency of molecules. The electron source contains three anodes, an electron collector, and a heater or filamentary cathode. In the placed in the gun electrodes A_1–A_3, made from polished nichrome plates with dimensions of 7 x 15 mm and thickness of 0.4 mm there are 1.5-mm-diameter holes that form a beam with a circular cross section. A 1-mm-wide gap between these electrodes is fixed with ceramic spacers. A side wall of the ionization chamber serves as the accelerating electrode A_3. The electron beam passes through the chamber wall that is opposite the electrode A_3 to the collector. The hole in the collector has been increased up to 6 mm for the electron flow to be detected correctly with allowance for the beam angular divergence and the influence on its trajectory of a potential that sags into the chamber and pulls ions out. The Faraday cup serves as the electron collector. Its depth and shape exclude an influence of the secondary electron emission on the results of the beam current measuring. The dimensions of the assembled electron gun with the collector are 15 x 30 x 8 mm.

The modernization of the electron source and its power supply resulted in the possibility to obtain electron beams with energies of 8–10 eV and energy spread $\Delta E_{1/2} < 0.5$ eV (FWHM) and control the electron energy with a computer, and thus automate the measurements of the ionized fragment yield.

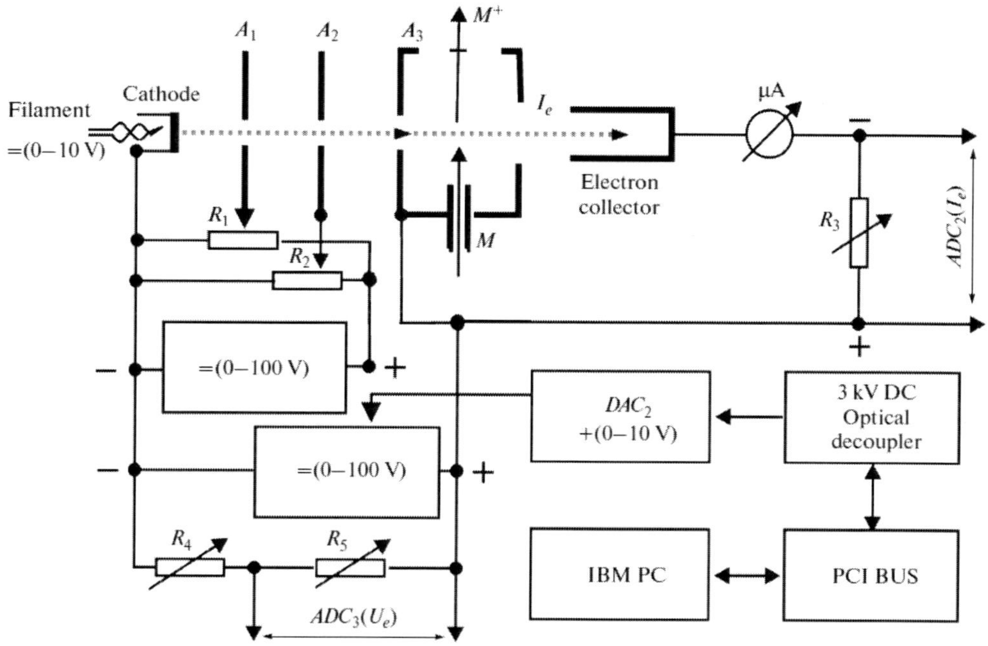

Figure 1. Schematic diagram of the power-supply system of the ion source electron gun.

The electron gun was powered from the high-stability DC sources. The potentials of the anodes A_1 and A_2 with respect to the cathode were controlled with potentiometers R_1 and R_2. The П4105 stabilizer was modified to provide a programmed control of its output by an external reference-voltage source served as the electron-accelerating voltage source. In our case, this is a PISO-DA2 12-bit digital-to-analog converter (DAC) [7]. The П4105 reference-voltage source is used in the manual control mode.

Resistor R_3 and a microampermeter are included in the circuit for measuring the electron currents I_e. The voltage drop across the resistor R_3 at a current of 100 μA is 10 V. The resistors R_4 and R_5 constitute a divider in the circuit for the measurement of the electron-accelerating voltage U_e. The voltage drop across resistor R_5 at an accelerating voltage of 100 V is 10 V. These voltages are converted into the digital codes by the 32-channel 12-bit ADC (PISO-813) [8].

The mass spectrometer ion source is kept at a high voltage (1– 5 kV) relative to the analyzer ground electrode, therefore, the measuring and controlling digital converters, voltmeters, microampermeters, and power-supply sources are insulated from the ground via the 3-kV dc optical decoupler.

The interconnections between the analog and digital controlling and recording modules and the computer are shown in Fig. 2 (some digital control devices are not shown in this figure). The electrometer for the ion current detecting and the magnetic-induction stabilizer of the analyzer for scanning the peak contour operate in the system for automatic measurements of the energy dependences of dissociative ionization cross sections for the molecule under study. In the ion-yield detection channel, the ADC_1 [9] converts the electrometer output voltage, which is proportional to the ion current I_i, into a digital code. The computer scans the peak contour and finds its intensity maximum through the converter DAC_1 [9]. The corresponding control voltage U_c is monitored with a B7-34A voltmeter. The computer holds this voltage U_c, which is the peak maximum at the input slit of the ion receiver during the entire measurement procedure, at the output of the induction stabilizer. The output voltage U_e of the P4105 controlled power supply is specified by computer instructions through the DAC_2.

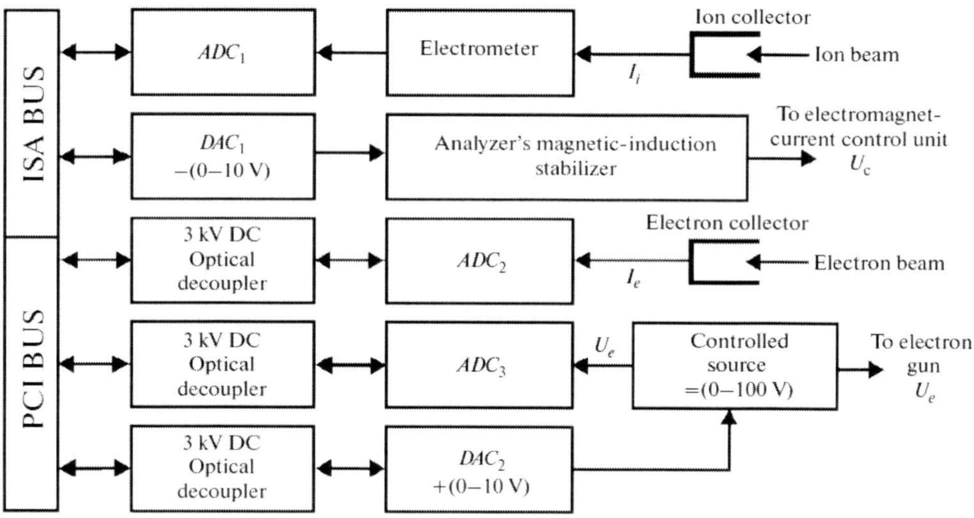

Figure 2. Schematic diagram of the system for scanning peaks and measuring the energy dependences of dissociative ionization cross the sections.

The effusion source of molecular beam is a thin-wall 15-mm-diameter stainless-steel vessel, into which the substance studied (in a form of a powder) is placed. The vessel is equipped with a vapor conduit, a ceramic tube 4 mm in the diameter and 10 mm in the length, which insulates it from the ionization chamber relative to a direct current. A molecular beam is formed by a 2-mm-diameter channel positioned directly at the chamber input. The necessary for the study molecular beam concentration is provided by thermal heating of the vessel using a tungsten heater supplied from a stabilized current source. The vessel temperature is measured by means of a calibrated chromel–alumel thermocouple.

Molecular flow *M* (see Fig. 1) to the ionization chamber is directed perpendicularly to the electron beam (the beams intersect in the plane of the figure). In the upper wall of the chamber there is a rectangular slit with the dimensions of 1 x 5 mm positioned in parallel to the electron-beam axis and serving for extracting ions.

The experimental procedure was as follows. First both electron and molecular beam sources were put into operation and after reaching the optimal conditions of beam generation a mass spectrum of the molecules under study was measured. Since the amino acid molecules may undergo strong thermal decomposition, the above mass spectra were measured serially at different molecular beam source oven temperatures. The temperature dependence of the ionized fragment production with m=30 a.m.u. (here m being the mass of the fragment) from the methionine molecule measured at the 66 eV ionization energy is shown in Fig. 3. When analyzing these data the conclusion can be drawn, that the temperature range of up to 120°C appears to be optimal to obtain a stable beam of the molecules under study, providing reliable detection of the signal at the ion detector and ensuring no essential effect of thermal degradation, i.e. the linear behavior of this dependence plotted in the semi-logarithmic scale proves the fact that in the temperature range under study the probability of production of the above fragment due to the thermal degradation of the parent molecule is negligible. This is confirmed by the data on the thermogravimetric studies of L-α amino acids [10].

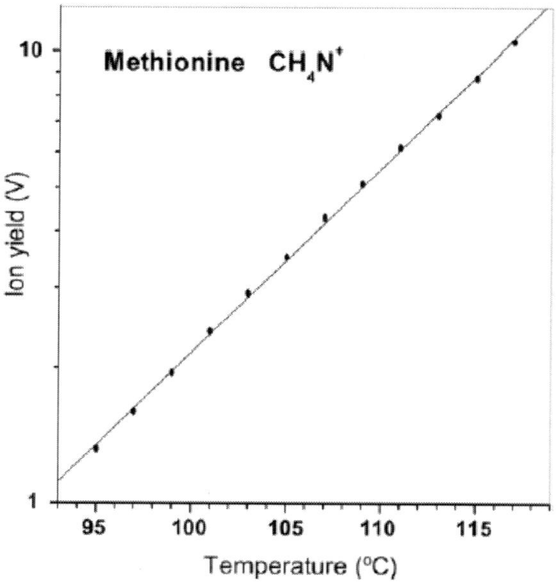

Figure 3. Temperature dependence of the CH_4N^+ ion yield from the methionine molecule.

The masses of the fragments produced were then fixed and the energy dependences of the relevant ion yields were measured. The energy was varied with an energy step of 0.1–0.3 eV, which enabled the threshold areas of the dissociative ionization functions for all the fragments under study to be measured. In this case the problem of the incident electron energy scale calibration becomes very significant. In order to calibrate the above scale we have measured first the threshold areas of the ionization functions for the two test gases – Ar and N_2. The experimental ionization thresholds were determined by means of a fitting technique suggested by T. Maerk's group (see, e.g., our earlier paper [11]) and used in the most of our previous studies. This technique is based on a least-square method approximation using the Marquardt-Levenberg algorithm. The details of this method can be found in a lot of scientific sources (see, e.g. [12]). The threshold areas of the energy dependences of the ionization cross-sections for Ar and N_2 are shown in Figs. 4a and 4b, respectively. For comparison, Figure 4a also shows the data on the production of Ar^+ ion measured by Freund et al.. [13]. One may note that the fitted curves have sharper thresholds as compared to the initial experimental ones. This is mainly due to the fact that fitting procedure does not take into account possible contributions of the excited ionic states lying just above the ionization threshold (note that fitting is carried out in a certain incident electron energy range that covers excitation thresholds for such states). However, this does not influence substantially the results obtained. As it is obvious from the figure presented, our data agree well with the Freund's results within the limits of the experimental uncertainty confirming the reliability of the fitting technique applied in this work. The absolute value of the Ar atom ionization threshold energy obtained by us (E=15.76 eV) agrees accurately with that known and included into a number of databases (see, e.g. [5]). The same is valid for the data obtained here for the nitrogen molecule (see Fig. 4b). The ionization threshold value for this molecule determined by us, i.e. 15.56 eV, agrees extremely well with the NIST data [5]. Thus, our technique obviously ensured reliable measurement of the absolute energy thresholds for the processes under study.

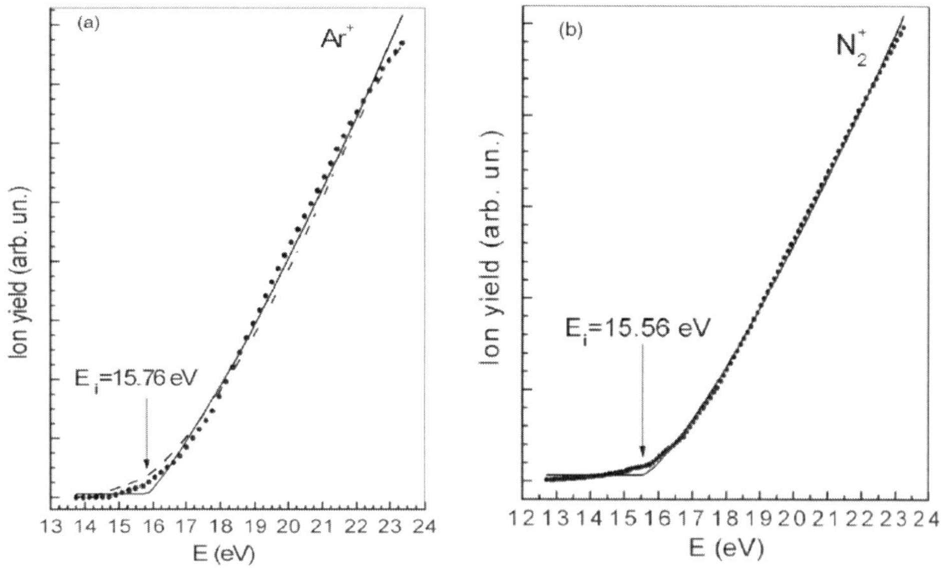

Figure 4. Near-threshold areas of the energy dependences of the ionization cross-sections for the Ar atom (a) and N_2 molecule (b). Solid circles – present experiment. Solid lines – our data fitted using the least-square technique. Dashed line – data of Freund et al.. [13]. Arrows indicate the ionization threshold values obtained using the fitting procedure.

Theoretical

The structure of the molecule and its fragments has been studied by the Becke's three-parameter hybrid functional, applying the non-local correlation provided by Lee, Yang, and Parr (B3LYP) [14], – a representative standard DFT method with the cc-pVTZ basis [15]. The structures of the molecule isomers and their fragments under study have been optimized with no symmetry constraint. The results of the analysis of the bond order and bond length of the methionine molecule isomers were used to find the weaker bonds that are most probably destructed. The vibration spectrum has also been evaluated to predict the possible elongation of bonds and the change of the angle with the aim to make decision on the most probable fragments produced by electron impact. In order to model the fragmentation processes, the possible fragment anions, cations and fragments with a zero charge have been evaluated both with and without geometry optimization to predict the influence of dissociation energy on the fragmentation processes. The appearance energies have been calculated as the difference between the total energy of the methionine molecule and the sum of the energies of the fragments predicted. Aiming to evaluate the above influence we have analyzed the following two cases: i) the single point energy calculation of the fragments was performed taking into account the geometry of a certain part of the methionine molecule (in these cases the fragments formation energy is not the lowest one); ii) the structure of the methionine molecule fragments has been optimized, i.e. the fragments were allowed to reach their equilibrium geometry and the obtained energy (the lowest energy of the fragments) has been used to calculate the dissociation energy. While applying this methodology, we have taken into account the processes when the molecular ions, formed with the energies just in excess of the ionization potential may not have the energy sufficient to be decomposed along the lowest energy pathway and will, thus, be collected and detected as such [16]. The GAMESS and the Gaussian program packages have been applied here [17, 18].

Results and Discussion

The initial area of the methionine molecule mass-spectrum measured in our experiment at the 70 eV electron energy (Fig. 5) is generally close to that published in the NIST database [5]. The methionine molecule image and the atom numbers used in this chapter are presented in Fig. 6.

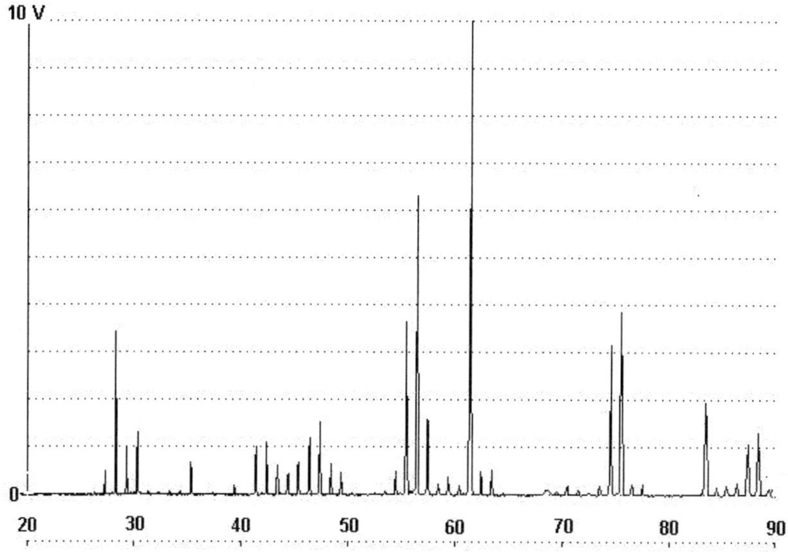

Figure 5. The initial area of the methionine molecule mass spectrum measured at the 70 eV electron energy.

The main peculiarity of the methionine molecule as compared to other amino acid molecules that impact the electron interaction process is the presence of a sulfur atom in the side chain being bound with the two carbon atoms.

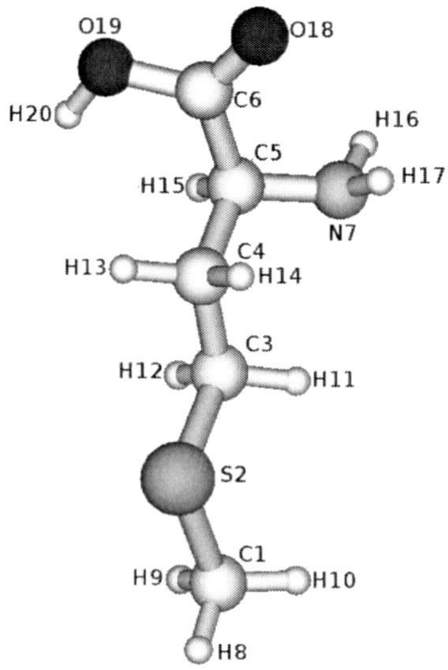

Figure 6. General view of the methionine molecule. The size and brightness of atoms are the same for all relevant figures.

Since the methionine molecule, besides the sulfur atom, comprises two isolated functional groups involving heteroatoms with lone pair electrons, while the carbonyl group comprises the π-electrons, its ionization could proceed on due to both electron loss from the non-bounding heteroatom *n*-orbitals and the elimination of the double-bond π-electron. However, from the thermodynamic viewpoint, in case of molecular ion dissociation, the charge should be located at the fragment with less ionization potential, i.e. in our case at the sulfur atom, since this atom possesses stronger electric donor properties as compared to the amine-, hydroxyl- and carbonyl- groups [19].

According to one of the methods of analyzing the regulations of organic molecules fragmentation based on the notions of charge localization in molecular ions the fragmentation most probably proceeds via the simple bonds nearest to the charge localization places, i.e. at the energies slightly exceeding the methionine molecule ionization potential the most probable is the C–S bond dissociation. At higher energies, when the electron energy is sufficient to ionize deeper molecular orbitals, the nitrogen and oxygen lone pair orbitals first of all, it will result in a dissociation of the neighboring bonds, producing, in general, more fragmentation channels.

In order to make quantitative estimation of the most probable methionine molecule dissociation channels we have calculated the ground-state molecule orbitals (MO) using the B3LYP/cc-pVTZ method to determine the atomic orbital contribution to MO. Table 1 presents the results of the MO calculation involving the valence atomic shell electrons only, since binding energies of core electrons are too large to influence the ion appearance energy. The ionization potential of the methionine molecule, according to the Koopman's theorem

[20], is 6.21 eV and is determined by a lone pair electron of the sulfur atom what coincides with the result presented.

Table 1. Description of energy, bond type and molecular orbitals per atomic orbitals with the largest contributions for the methionine molecule calculated by the B3LYP/cc-pVTZ method. The largest molecular orbital coefficients, the atom and the atomic orbital types of this atom are presented.

Energy, eV	Assignment	Description of the orbital per atomic orbital
-14.96	σ_p	-0.18 (2pz,C6)+ 0.16 (2py, N7)
-14.50	σ_p	0.27(2px,C1)
-13.31	π	0.13(2px, C4)+0.16(2py,C4)+N7(-0.19(2py,N7)-0.09(2px,N7)-0.08(2py,O18)-0.098(2pz,O18)
-12.62	π	-0.202 (2py,C3)-0.17 (2px,O18) +0.18 (2pz,O18)
-12.57	σ_p	0.22(2px,C6)+0.13(2py,C6)-0.12(2px,N7)-0.11(2py,N7)-0,20 (2px,O18)+0.21(2px,O19)
-12.08	σ_p	-0.24(2px,C1)+0.15 (2py,O18)+0.12(2pz,O18)
-11.87	π	-0.21(2px,C1)+0.23(2py,C1) -0.13 (2py,O18)
-11.70	π	0.13(2px,C1)+0.25(2py,C1)+0.13 (2py,O18)
-11.32	σ_p	-0.12 (2px,C3)-0.18(2pz,C3)+0.19(2pz,C4)-0.14(2px,O18)-0.16(2px,O19)
-10.87	σ_p	0.11(2px,O18)+0.38(2pz,O19)
-10.59	σ_p	0.22(2pz,C3)-0.18(2px,C4)-0.20(2py,C4)+0.10(2py,O18)
-10.17	π	-0.20(2py,C3)+0.26(2py,C4)
-9.70	σ_p	-0.22(2pz,C1)-0.12(2pz,C3)+0.13(2px,C3)-0.16(2pz,S2)+0.12(2px,O19)
-9.04	π	-0.33(2px,O18)+0.37 (2px,O19)
-8.52	σ_p	-0.15(2pz,C1)-0.09(2px,C1)-0.195(2px,S2)-0.12(2pz,S2)-0.20 (2px,C3)
-8.04	π	0.24(2py,O18)-0.37(2pz,O18)+0.11(2py,O19)
-6.94	π	(-0.309 (2pz))
-6.21	π	HOMO 0.29 (2py,S2)

It should be noted that the electron binding energies calculated by us differ from those presented in [21]. These differences are due to the different approaches applied for the evaluation of the electron binding energy. In [21], electron binding energies are determined within the outer-valence Green function OVGF method [22], which takes into account both the MO reorganization energy and electronic correlation through the expansion series associated with each MO energy, i.e. with each primary ionization event. The method has been recognized as providing very satisfactory results and proved that the ionized band was not related to the shake-up ionized state [23]. However, the above method is expensive either in the hard disk volume and/or in the CPU time and usually is adopted in case when electron binding energies experimentally measured and calculated are compared. Thus, to avoid huge hard disk volume and the CPU time consuming, the ionization energies were calculated from the B3LYP/cc-pVTZ orbital energies, that are usually obtained with the errors much larger than 2 eV. Therefore, we prefer to calculate the first ionization potential as the difference between the energy of both the molecule and its ion, while the results presented in Table 1 are shown only for orientation aiming to foresee the charge localization and determine, which orbital could be assigned to the specific fragmentation channel.

The total (binding) energy calculated when the geometry optimization of the methionine molecule is performed is -800.27 a.u. and the total (binding) energy of cation is -799.976 a.u., so the ionization energy calculated as the difference in energies between the D methionine cation and neutral molecule is 0.30 a.u. or 8.09 eV. These values coincide excellently with the data of Dehareng and Dive [23] who calculated the vertical ionization energies for three methoinine conformers and one of the conformers had a first ionization energy of 8.09 eV.

It is generally known that ionization energy represents the absolute minimum energy required to ionize the neutral particle concerned. The further fate of the molecular ion depends on the shape of its potential energy surface. If its potential energy is minimal and the level of its excitation is below the energy barrier for dissociation, the ion may exist for a very long time. Ions having the internal energy above the dissociation energy level will dissociate at some point leading to the appearance of fragment ions in a mass spectrum [24].

The intensity of the aliphatic amine molecular ion peaks in the mass spectra is very low due to the high probability of decay processes initiated by the cation-radical center localized on the amine group. However, methionine displays a rather different pattern. Firstly, because the sulfur atom lone pair is the HOMO, therefore, at low energies it is ionized more preferably than the nitrogen lone pair. Secondly, the sulfide group stabilizes the positive charge and reduces fragmentation [25]. In our mass-spectrum (see Fig. 5), the initial molecular ion peak has relative intensity about 20%, while the NIST data [5] give that from 13.7 to 28.0%, and this value exceeds greatly the share of the similar molecular ions in the other aliphatic amino acid mass spectra. Thus, ionization of a sulfur lone pair orbital evidently causes much less molecular fragmentation. The mass-spectrum itself is characterized by a low selectivity what indicates the lack of a dominant decay pathway for the molecular ion.

Table 2. Calculated methionine molecule bond lengths and orders.

Bonds	Bond length, Å Neutral molecule	Bond order Neutral molecule	Bond length, Å Ionized molecule	Bond order Ionized molecule
C1-S2	1.886	0.938	1.886	0.903
C1-H8	1.087	0.935	1.085	0.910
C1-H9	1.087	0.924	1.087	0.902
C1-H10	1.087	0.924	1.087	0.897
S2-C3	1.899	1.024	1.877	0.981
C3-C4	1.525	0.886	1.531	0.809
C3-H11	1.091	0.897	1.094	0.873
C3-H12	1.086	0.917	1.088	0.892
C4-C5	1.553	0.906	1.576	0.865
C4-H13	1.093	0.947	1.091	0.934
C4-H14	1.093	0.936	1.091	0.915
C5-C6	1.539	0.854	1.566	0.851
C5-N7	1.456	1.008	1.429	0.976
C5-H15	1.096	0.925	1.090	0.913
C6-O18	1.222	1.969	1.216	1.927
C6-O19	1.386	0.951	1.359	0.999
N7-H16	1.012	0.876	1.009	0.844
N7-H17	1.014	0.869	1.013	0.825
O19-H20	0.978	0.805	0.978	0.788

It is known from the quantum-chemical calculations on determination of the geometry and interatomic distances of the optimized molecular structures for different organic molecules and their single-charged ions that the distances between the skeleton chain carbon atoms after ionization are resized (see, e.g. [26]). The electron loss by the molecule results in the fact that some skeleton bonds elongate, whereas the other bonds become a bit shorter. In this case the calculated value of the energy required for the C–C bond dissociation in the ionized molecules, depending on the fragmentation channels, is 0.84 to 3.45 eV [27].

Table 3. Dihedral angles between bonds and angles between planes.

Bonds	Neutral molecule	Ionized molecule
	Angles, degrees	Angles, degrees
C1-S2-C3	98.62	102.83
S2-C3-C4	109.32	108.45
C3-C4-C5	111.87	109.46
C4-C5-N6	110.70	112.17
C4-C5-C7	109.43	105.57
C5-N7-O8	116.46	117.39
C5-N7-O9	124.26	120.66
S2-C1-H10	106.45	106.141
S2-C1-H11	110.26	109.29
S2-C1-H12	110.09	109.09
S2-C3-H13	108.65	117.87
S2-C1-H14	107.84	108.49
C3-C4-H15	109.80	111.72
C3-C4-H16	109.90	110.70
C4-C5-H17	108.20	109.43
C5-N6-H18	114.20	120.94
C5-N6-H19	113.13	118.27
C7-O8-H20	113.89	117.01
C1-S2-C3-C4	-175.94	124.101
S2-C3-C4-C5	178.88	172.21
C3-C4-C5-N6	-61.96	-56.43
C3-C4-C5-C7	173.02	-175.63
C4-C5-N7-O8	-73.30	-99.10
C4-C5-N7-O9	105.98	78.86
C3-S2-C1-H10	179.82	-176.71
C3-S2-C1-H11	-61.31	-57.59
C3-S2-C1-H12	61.00	64.05
C1-S2-C3-H13	-55.30	107.27
C1-S2-C1-H14	63.01	0.48
S2-C3-C4-H15	-58.74	-66.51
S2-C3-C4-H16	58.28	53.75
C3-C4-C5-H17	55.64	64.38
C4-C5-N6-H18	163.96	126.72
C4-C5-N6-H19	-68.41	-76.77
C5-C7-O8-H20	-0.86	-6.78

To estimate the change in the methionine molecule geometry as the result of ionization, we have calculated both the bond lengths and the bond orders of the neutral and ionized methionine molecules after the geometry optimization.

Having taken into consideration the results and implementing the Mulliken and Lowdin population analysis data [28], the weakest bonds in the methionine molecule were determined. The bond lengths calculated and bond orders of the neutral and ionized methionine molecule are listed in Table 2, while the changes in the geometry are shown in Fig. 7 and Table 3.

According to the results obtained for the neutral methionine molecule, the weakest bonds are C3–C4, C5–C6 and C4–C5. As a result of ionization, these bonds elongate, i.e. the probability of the molecular ion dissociation along these bonds increases. On the other hand, ionization of the molecule leads to the shortening of the carbon-heteroatom bonds, especially the C5–N7 bond, so deamination becomes improbable.

The highest peak in the methionine molecule mass-spectrum (Fig. 5) corresponds to the $C_2H_5S^+$ ion (*m*=61 a.m.u.), which is complementary to the $C_3H_6NO_2^+$ ion (*m*=88 a.m.u.). This pair of fragments is formed at a simple break of the weakest C3–C4 bond of the initial molecule, and the charge is mainly localized at the fragment comprising the sulfur atom. In this case the sulfur atom addiction to ionization and decay initiation are clearly seen. Inclusion of any possible cases of charge distribution at this bond dissociation results in the production of the C_2H_5S and $C_3H_6NO_2$ fragments according to the following schemes:

$$C_5H_{11}NO_2S + e \rightarrow \begin{cases} C_2H_5S^+ + C_3H_6NO_2^{\,-} + e & (1) \\ C_2H_5S^+ + C_3H_6NO_2^{\,0} + 2e & (2) \\ C_2H_5S^{\,-} + C_3H_6NO_2^{\,+} + e & (3) \\ C_2H_5S^{\,0} + C_3H_6NO_2^{\,+} + 2e & (4) \end{cases}$$

The calculated appearance energies for the above fragments are listed in Table 4. The appearance energy E_{ap} was calculated as: $E_{ap} = |E_{meth}| - |\Sigma(E_i)|$. Here E_{meth} is the total energy of the neutral methionine molecule, while E_i is the total energy of all fragments produced. This calculation does not take into account the activation energy of the molecular ion fragmentation.

The most energetically favorable channels of the above cations formation correspond to pathways (1) and (3). Comparison of the intensities of the peaks leads to the prediction that formation of the $C_2H_5S^+$ cation is more probable than that of $C_3H_6NO_2^+$, although the opposite conclusion could be made from the result of the appearance energies calculated (Table 4). Thus, the binding energies per atom for these cations were calculated as well, and the results obtained prove that the above $C_2H_5S^+$ cation appears to be more stable than $C_3H_6NO_2^+$. It could be the main reason for that in the mass spectrum the peak at *m*=61 a.m.u. is more intense than that at *m*=88 a.m.u. The complementary ions $C_2H_5S^+$ and $C_3H_6NO_2^+$ with low intensities were observed in the photoionization spectra at the 8.43 eV photon energy. At 11.62 eV their intensities are almost the same, while starting with 16.67 eV the peak at *m*=61 a.m.u. corresponding to the $C_2H_5S^+$ ion dominates [25].

The analysis of both the electron impact (EI) and photoionization (PI) mass spectra shows that the main channel of the methionine molecule fragmentation above 16 eV is related to the rupture of the β-bond with respect to the sulfur atom. However, the appearance energy for this ion experimentally obtained in this work (12.4±0.1 eV) (see also [29]) allows us to conclude that this channel is not related only to the removal of a lone-pair electron from the sulfur atom. To make this ionization channel dominant, the ionization of at least of the third-highest occupied MO is required with further charge stabilization on the sulfur atom.

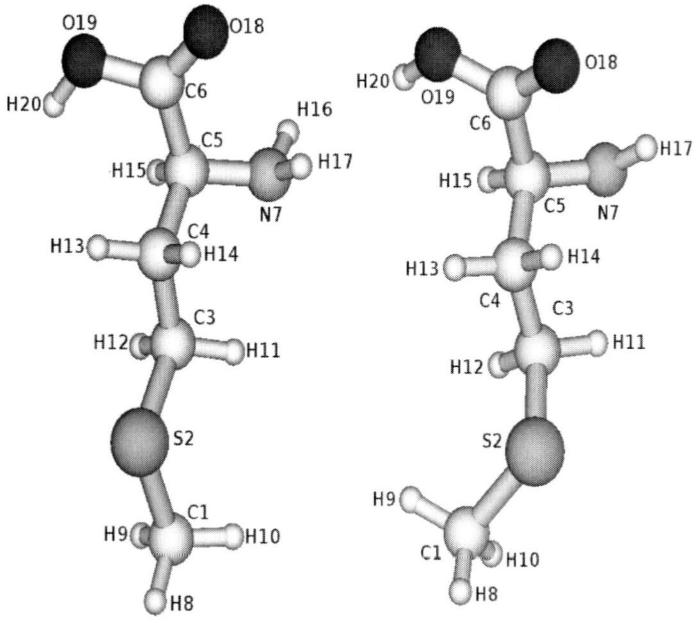

Figure 7. Neutral (left) and ionized (right) methionine molecule. The number of atoms in both figures remained the same.

Table 4. Calculated appearance energies (eV) for the C_2H_5S and $C_3H_6NO_2$ fragments.

C_2H_5S (m=61 a.m.u.) charge	$C_3H_6NO_2$ (m=88 a.m.u.) charge	Methionine molecule geometry not changed*	Methionine molecule geometry changed**
1	-1	13.71	7.17
1	0	14.53	10.71
-1	1	13.49	8.43
0	1	14.29	8.72

* 'Methionine molecule geometry not changed' means that the single point energy calculation of the fragments taking into account the geometry of the certain part of the methionine molecule was performed.

** 'Methionine molecule geometry changed' indicates that the equilibrium geometry structure of the methionine molecule fragments is investigated.

The next prominent peak in the mass-spectrum of the methionine molecule is due to the formation of the parent molecule fragment $C_3H_6N^+$ with the m=56 a.m.u. mass. This fragment may conceivably arise due to elimination of the CH_3SH and COOH fragments. In the PI and EI spectra, the peaks with the m=104 a.m.u. and m=101 a.m.u. masses corresponding to (M^+–

COOH) and (M^+–CH_3SH) are present (here M^+ is a molecular ion produced at the initial molecule ionization). The intensities of these peaks in the PI spectra below 11.62 eV exceed the intensity of the peak of the *m*=56 a.m.u. fragment. At higher photon energies (>16 eV), the intensity inversion of these peaks is observed. It should be noted that at photon energies below 16 eV the efficiency of the PI channels, related to the rupture of the C3–S2 and C5–C6 bonds, is higher than that of the channel related to dissociation of the weakest C3–C4 bond leading to the $C_2H_5S^+$ fragment formation. Thus, the removal of the n_S-electron (or the HOMO-ionization) results in the appearance of fragmentation channels related to the dissociation of the C3–S2 and C5–C6 bonds. Obviously, the positive charge in this case is localized not on the sulfur but on the nitrogen atom.

The CH_3S+H and COOH fragments could be removed at the same time or it may be the two-step process. Some relevant results are presented in Tables 5 and 6. The lowest energy is reached when the CH_3S + H fragments are joined to the CH_4S molecule. Theoretical investigations prove that the removal of the H13 atom requires the smallest amount of energy, thus the CH_4S molecule is produced due to the H13 atom migration to the sulfur atom via the 4-term transient state.

Table 5. Calculated appearance energies (eV) for the C_3H_6N and CH_4S + COOH fragments of the neutral methionine molecule.

C_3H_6N, (m=56 a.m.u.) charge	CH_4S+COOH, charge	Methionine molecule geometry not changed *	Methionine molecule geometry changed **
1	-1	9.16	1.24
1	0	9.63	4.38

*,** – see explanatory notes in Table 4.

Table 6. Calculated energies (in eV) required for the C_3H_6N and CH_4S fragment formation from the methionine molecule with the loss of the COOH fragment.

C_3H_6N (m=56 a.m.u.) charge	CH_4S charge	(M^+-COOH) geometry not changed *	(M^+-COOH) geometry changed **
1	-1	8.06	2.93
1	0	10.04	1.73
1	1	18.86	11.16

*,** – see explanatory notes in Table 4.

The appearance energy for the $C_3H_6N^+$ ion, measured in our experiment is 11.2±0.1 eV and it is higher than the calculated one. So, the process of simultaneous formation of the (CH_4S + COOH) and $C_3H_6N^+$ fragments from the neutral methionine molecule under EI is not realistic. Thus, the calculations of the energies required for the $C_3H_6N^+$ and CH_4S fragments formation from the methionine molecule after the loss of the CO_2H fragment, were carried out (Table 6).

In fact, Table 6 presents the excess energy of the $C_4H_{10}NS^+$ ion required for its subsequent fragmentation with the production of the $C_3H_6N^+$ ion. The experimental difference in the appearance energies E_{ap} for the *m*=104 a.m.u. and *m*=56 a.m.u. fragments measured in this work (as well as in [29]) is 1.5±0.2 eV, thus, the general scheme for the cascade process is as follows:

$$C_5H_{11}NSO_2 + e \rightarrow COOH^0 + C_4H_{10}NS^+ + 2e$$
$$\searrow C_3H_6N^+ + CH_4S^0. \qquad (5)$$

It should be noted that the geometrical and electronic structures of the C_3H_6N fragment could be different because of the removal of different H atoms. Our calculations show that the removal of the H13 atom requires the smallest amount of energy. From the results of the comparison of the appearance energies as well as from the cation energy itself, we predict that the positively charged ion with the m=56 a.m.u. mass has the form presented in Fig. 8.

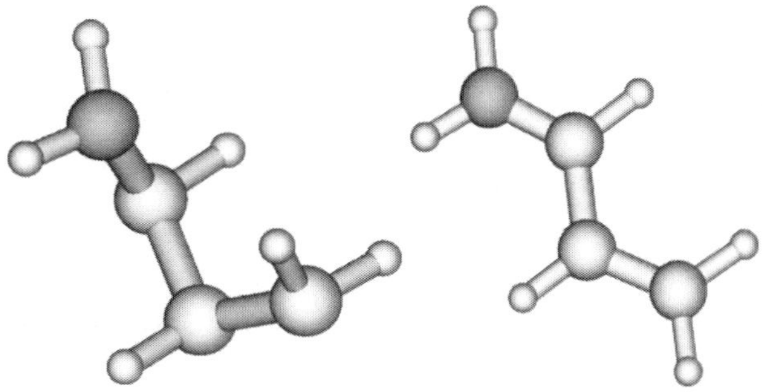

Figure 8. View of the m=56 a.m.u. cation when the methionine molecule geometry is not changed (left) and when the equilibrium geometry of the cation is reached (right).

In this case, the C3–C4–C5 atoms are located on one plane at the angle of 120.57°, i.e. the hybridization of the carbon atoms is changed: $sp^3 \rightarrow sp^2$, and the pseudoconjugation system is formed involving a single unsaturated moiety and an atom with a lone pair of electrons in the p-orbital. Electron delocalization occurs over three atoms.

The next prominent peak in the mass-spectrum of the methionine molecule is due to the formation of the fragment with the mass of m=30 a.m.u. Such a peak is characteristic for alkyl amines and usually dominates in the mass-spectra of the relevant compounds.

On the basis of the results of our investigations, it was concluded that the assignment of this peak is related to the formation of the CH_4N^+ fragment [29]. It should be mentioned that formation of the CH_4N^+ fragment from the alkyl amines with side chains is complicated because not only the electron-impact induces the bond cleavage, but also the rearrangement of one hydrogen atom upon fragmentation is required.

The CH_4N^+ fragment could be formed according to the following schemes:

$$C_5H_{11}NO_2S + e \rightarrow \begin{cases} (CH_3N+H)^+ + (COOH+C_3H_6S)^- + e & (6) \\ (CH_3N+H)^+ + (COO+C_3H_7S)^- + e & (7) \\ (CH_3N+H)^+ + (COO+C_3H_7S)^0 + 2e & (8) \\ (CH_3N+H)^+ + (COOH+C_3H_6S)^0 + 2e & (9) \end{cases}$$

The hydrogen atom attachment to CH_3N^+ is followed by the energy release of 6.12 or 7.85 eV respectively, when H originates from the C_3H_7S or COOH fragment, respectively. Thus, the hydrogen atom from the hydroxyl group takes part in the CH_4N^+ fragment formation. In this case, the lowest energy is required to divide the methionine molecule into the $(COO + C_3H_7S)^-$ anion and the CH_4N^+ cation.

The fragment peak CH_4N^+ (m=30 a.m.u.) in the methionine molecule mass spectrum is accompanied by the satellite peaks with the masses of m=28 a.m.u. and m= 29 a.m.u.

The fragment with the mass of m=29 may be assigned to the COH^+ or CH_3N^+ ion. According to the results of theoretical investigation, it is possible to state that in the case of methionine the COH fragment could not be formed because there is no possibility to form the $C_4H_{10}NS$ + O compound, i.e. the probability of the simultaneous C–O and C–C bonds rupture is very low. Hence, the fragment with m=29 a.m.u. could be the CH_3N^+ cation. The lowest appearance energy for this fragment is required when the methionine molecule is divided into the ion pair according to the following scheme (when the initial geometry of the methionine molecule is not changed):

$$C_5H_{11}NO_2S + e \rightarrow CH_3N^+ + (COOH{+}C_3H_7S)^- + e\,. \qquad (10)$$

Despite the fact that production of the CH_3N^+ fragment takes place at the minimal number of structural changes as compared to the production of the ions with m=30 a.m.u. and m=28 a.m.u., the intensity of the peak with m=29 a.m.u. is lower as compared to that with m=30 a.m.u. or m=28 a.m.u. It should be noted that the calculated appearance energy for the CH_3N^+ fragment is at least 3.48 eV higher than that for the $(CH_3N + H)^+$ fragment. On the basis of these results we predict that the electron-impact fragmentation of methionine producing the $(CH_3N + H)^+$ fragment is more probable than that for the CH_3N^+ fragment.

The ion with the m=28 a.m.u. mass may have the following gross formulae: CH_2N, CO and C_2H_4. The investigations of the simplest amino acid – glycine [30] suggested that this peak is due to the CH_2N^+ and/or CO^+ ions. However, the mass-spectra of the deprotonated glycine (with the $HNCD_2CO_2H$ and $H^{15}NCH_2CO_2H$ composition) show that this fragment contains a nitrogen atom, while the experiment for the deuterated d5- and d3- glycine [31] has unambiguously shown that this peak belongs to the CH_2N^+ (CD_2N^+) ion. The comparison of the stability of the positively charged ions CH_2N^+ and CO^+ (i.e. the binding energy per atom in eV), calculated by us also leads to the conclusion that the formation of the CH_2N^+ cation is more probable than that of the CO^+ cation.

Comparison of the values of the binding energy per atom indicates that the CO^+ ion is less stable than the CH_2N^+ and $C_2H_4^+$ ones. Hence, observation of the CH_2N^+ and $C_2H_4^+$ fragments due to the electron impact is more probable than that for CO^+. Evidently the CO^+ fragment could be produced according to the following schemes:

$$C_5H_{11}NO_2S + e \rightarrow CO^+ + (HO + C_4H_{10}NS)^- + e; \qquad (11)$$

$$C_5H_{11}NO_2S + e \rightarrow CO^+ + (HO + C_4H_{10}NS)^0 + 2e. \qquad (12)$$

The calculated appearance energies are presented in Table 7.

Table 7. Calculated appearance energies (in eV) for the CO and (HO + $C_4H_{10}NS$) fragments of the methionine molecule.

CO fragment, charge	(HO+$C_4H_{10}NS$) fragment, charge	Geometry not changed**	Geometry changed**
1	-1	17.66	15.53
1	0	18.50	14.92

*,** – see explanatory notes in Table 4.

It is clearly seen that the appearance energy of the CO^+ fragment according to the schemes given above is higher than the measured energy value (11.9±0.1 eV). The results described make it possible to predict that the positively charged fragment with mass m=28 could not be the CO^+ ion.

Hence, two ways of formation of the CH_2N fragment were investigated because this fragment could be produced when two C–C and C–H or N–H bonds of the methionine are ruptured. It implies that the CH_2N fragment could have two structures – HCNH or CH_2N.

The structure of the CH_2N^+ ion depends on the parent or intermediate ion bonds being broken. Four possible relevant isomers are shown in Fig. 9. It is important that in cases when different bonds of the CH_4N^+ or CH_3N^+ cation are broken the fragments become planar after the geometry optimization. Calculations [31] have shown that the HCNH structure is the most stable of those presented in Fig. 9. According to our calculations, after the geometry optimization with the B3LYP cc-pVTZ approach application, the trans- and cis-isomers (III and IV in Fig. 9) transit to the most stable structure with the linear configuration (point group C_s). Note that the C and N atoms in this case undergo the *sp*-hybridization.

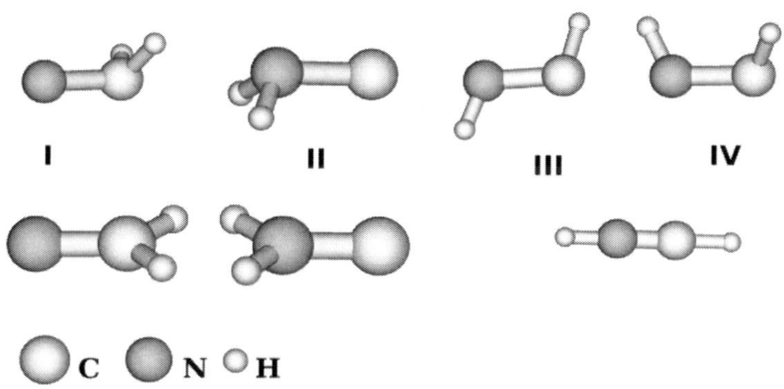

Figure 9. CH_2N^+ cation isomers before (top) and after (bottom) the geometry optimization.

The analysis of the appearance energies for the CH_2N^+ fragment produced from the neutral and ionized molecule shows that the calculated appearance energy of 11.83 eV according to the scheme

$$C_5H_{11}NO_2S^+ \rightarrow CH_2N^+ + (CHO_2 + H + C_3H_7S) \quad (13)$$

coincides with the measured value (11.4±0.1 eV). However, in the near-threshold energy region production of the CH_2N^+ ion may be caused by some other mechanisms.

The positively charged CH_3N^+ and CH_2N^+ ions could result from the secondary dissociation due to deprotonation of the CH_4N^+ ion, while the CH_2N^+ ion could also appear due to deprotonation of the CH_3N^+ ion. The above secondary dissociation is more possible in case when the energy, transferred to the initial molecule, increases.

The methionine molecule mass spectrum reveals a diffuse peak at the mass of about $m^*\sim 26.1$ a.m.u. (Fig. 10, the semi-logarithmic scale should be noted here) corresponding to the 30→28 transition with the detachment of a neutral fragment with $m=2$ a.m.u., i.e. the secondary fragmentation of the CH_4N^+ ion occurs. Thus, we have experimentally observed dehydration that proceeds according to the scheme $CH_4N^+ \rightarrow (H + H)^0 + CH_2N^+$ producing the molecular hydrogen.

Taking into account the measured appearance energy values for the CNH_2^+ and CNH_4^+ fragments (11.4±0.1 and 11.0±0.1 eV, respectively) and those calculated for these ions produced from the initial molecule, one may assume that in the near-threshold energy region the main mechanism of the CNH_2^+ ion production is a cascade dissociation via the CNH_4^+ ion formation:

$$C_5H_{11}NO_2S^+ \rightarrow CNH_4^+ + C_4H_7O_2S^0 \rightarrow CNH_2^+ + H_2 + C_4H_7O_2S^0 \quad (14)$$

Due to this dehydration, when the stable hydrogen molecule is formed as the dissociation product, the fragment with the mass of $m=29$ a.m.u. is difficult to be produced as well. Such a pattern of the peak intensity distribution is an argument in favor of the hypothesis that the driving force of the molecular ion decay is not the charge localization at the heteroatom with the further break of the closest to it bonds, but the stability of the dissociation products.

We assign the peak at $m=45$ a.m.u. to the $COOH^+$ fragment because in order to produce the above fragment the rupture of only the C5–C6 bond is required, while formation of other isobaric fragments, such as CHS^+ and $C_2H_7N^+$, require a break of several bonds and, in some cases, the rearrangement of the hydrogen atoms. On the other hand, it should be noted, that the loss of COOH, NH_2 and $COO+NH_3$ together with formation of $COOH^+$ are the major channels of the methionine molecule dissociative ionization [32]. Different processes of the $COOH^+$ fragment formation were analyzed:

$$C_5H_{11}NO_2S + e \rightarrow \begin{cases} CHO_2^- + C_4H_{10}NS^+ + e & (15) \\ CHO_2^0 + C_4H_{10}NS^+ + 2e; & (16) \\ CHO_2^+ + C_4H_{10}NS^- + e & (17) \\ CHO_2^+ + C_4H_{10}NS^0 + 2e & (18) \end{cases}$$

It is possible to notice that the lowest energy required to divide the methionine molecule aiming to form the CHO_2^+ cation corresponds to scheme (16). The experimentally measured appearance threshold for this fragment is 13.5±0.1 eV. The comparison of this value with the data from Table 8 shows that the production of the CHO_2^+ fragment may proceed via channels (17) and (18).

It should be noted that, according to schemes (15–18), the fragment with the $m=104$ a.m.u. mass is formed. By comparing the calculated and measured (9.7±0.1 eV) appearance energies for this positively charged ion, the conclusion can be made that dissociation with the $C_4H_{10}NS^+$ ion production from the parent ion proceeds according to pathways (15) and (16).

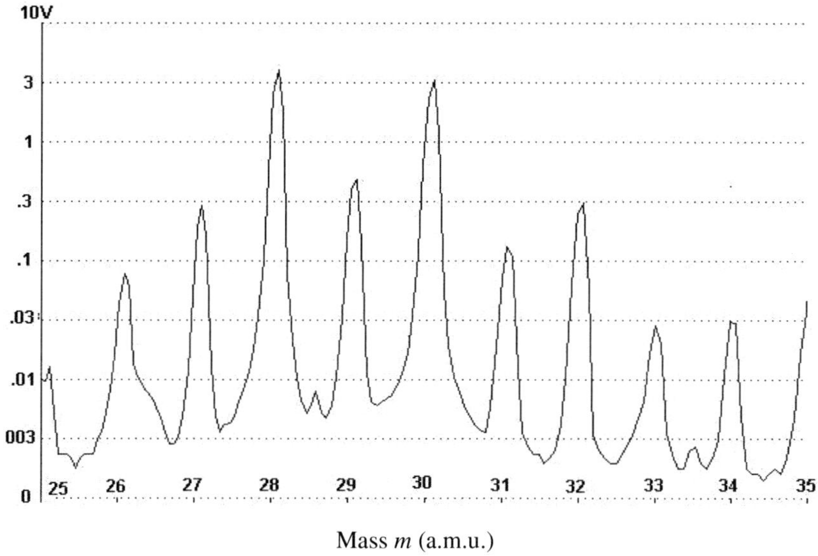

Figure 10. Low-mass area of the methionine molecule mass spectrum measured at the 70 eV electron energy.

Table 8. Calculated appearance energies (in eV) for the CHO_2 and $C_4H_{10}NS$ fragments.

CHO_2 (m=45 a.m.u.), charge	$C_4H_{10}NS$ (m=104 a.m.u.), charge	Methionine molecule geometry not changed*	Methionine molecule geometry changed**
-1	1	10.01	7.35
0	1	10.77	8.7
1	-1	14.98	12.34
1	0	14.24	11.76

*,** – see explanatory notes in Table 4.

Since the COOH and $C_4H_{10}NS$ fragments are complementary, the intensity of the corresponding peaks in the mass-spectrum may characterize the efficiency of the above reaction channels. According to our data, the intensity of the peak at m=104 a.m.u. is higher than that of the peak at m=45 a.m.u., thus, in case of the C5–C6 bond dissociation the cation center is mainly displaced to the $C_4H_{10}NS$ fragment.

The mechanism of the $C_4H_{10}NS^+$ fragment formation requires the C5–C6 bond of the molecular ion to be broken. This could be visualized as the β-decay towards the nitrogen atom producing the immonic structure typical in the dissociative ionization of amines, where the nitrogen atom is capable of the effective stabilization of the positive charge:

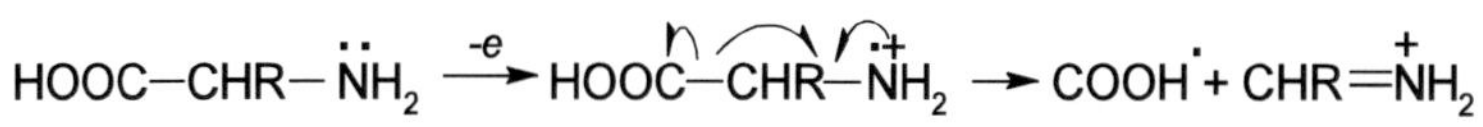

(19)

Here R is an amino acid side chain. In the case of methionine R is C_3H_7S. The one-electron shift (⌒) from the ionized nitrogen atom and α-carbon atom results in formation of a double bond accompanied by energy release. The peak at *m*=46 a.m.u. in the mass-spectrum of the methionine molecule is assigned to the $CH_2O_2^+$ ion.

The above ionized fragments may be produced according to the following schemes:

$$C_5H_{11}NO_2S + e \rightarrow \begin{cases} (H+CHO_2)^+ + C_4H_9NS^0 + 2e \rightarrow CH_2O_2^+ + C_4H_9NS^0 + 2e & (20) \\ (H+CHO_2)^+ + C_4H_9NS^- + e \rightarrow CH_2O_2^+ + C_4H_9NS^- + e & (21) \end{cases}$$

The calculated appearance energy of the ion when the neutral $C_4H_9NS^0$ is formed is equal to 13.22 eV, while in the case of $C_4H_9NS^-$ formation this energy is equal to 12.59 eV. These two values are close to the measured appearance energy (13.1 ± 0.1 eV). Thus, we predict that both reactions are possible. It should be emphasized, that several different possibilities to form the ion with *m*=46 a.m.u. were also investigated. However, only in one case the calculated and measured appearance energy values fit very well. This is the case when methionine looses the COOH group and the H13 atom joins CHO_2 to form $CH_2O_2^+$.

The peaks at *m*=74 a.m.u. and *m*=75 a.m.u. are present in the mass-spectrum with almost equal intensities. At first sight, these two fragments are complementary and are formed due to the C4–C5 bond rupture. However, the satellite peaks at *m*=76 a.m.u. and *m*=77 a.m.u. with the intensities of 7.47% and 4.47%, respectively, considering the contribution of ^{13}C to *m*=76 a.m.u., allow us to state that both fragments with the *m*=74 and *m*=75 a.m.u. masses include the sulfur atom, thus confirming the statement of Junk and Svec [33]. The fragment with *m*=75 a.m.u. (i.e. $C_3H_7S^+$) is produced due to the C4–C5 bond dissociation, whereas the fragment at *m*=74 a.m.u. (i.e. $C_3H_6S^+$) is formed as a result of the hydrogen atom migration in the transient complex with the subsequent dissociation of the skeleton C4–C5 bond. In this case the neutral fragment is detached and its structure corresponds to the glycine molecule.

Calculation of the appearance energies for the $C_2H_4NO_2^+$ (*m*=74 a.m.u.) and $C_3H_7S^+$ (*m*=75 a.m.u.) ions at the C4–C5 bond dissociation shows that the energy minimum corresponds to the formation of an ion pair (Table 9). Comparison of the calculated appearance energies for these fragments allows us to conclude that the process $C_4H_{11}NSO_2$+e → $C_2H_4NO_2^-$ + $C_3H_7S^+$ +e is more probable.

Table 9. Calculated appearance energies (eV) for the $C_2H_4NO_2$ and C_3H_7S fragments.

$C_2H_4NO_2$ (*m*=74 a.m.u.) charge	C_3H_7S (*m*=75 a.m.u.) charge	Methionine molecule geometry not changed*	Methionine molecule geometry changed**
-1	1	11.83	8.07
0	1	13.12	8.96
1	-1	12.98	7.85
1	0	13.46	9.79

*,** – see explanatory notes in Table 4.

The peak at *m*=83 a.m.u. is present in the mass spectrum with a moderate intensity. On the basis of a simple analysis (mass calculation of all possible fragments formed due to any bond rupture) we may predict that the fragment with m=83 a.m.u. is the product of several short-run processes. In [25], it was assumed that this fragment is produced due to a cascade fragmentation of the methionine molecule via the formation of the ion with the m=131 a.m.u. mass (due to the loss of H_2O), and the further loss of the CH_3SH neutral fragment. The $C_4H_5ON^+$ fragment may have either the cyclic structure of the amine-derived furan or the linear one (Fig. 11). It is necessary to note, that the $C_4H_5ON^+$ cation is divided into the C_3H_5N and CO fragments at the equilibrium point. Meanwhile the mechanism of production of this fragment and its structure is not experimentally established. The minimal number of structural transformations corresponds to elimination of H_2O with hydrogen transfer to the sulfur atom and with the C3–S2 bond rupture.

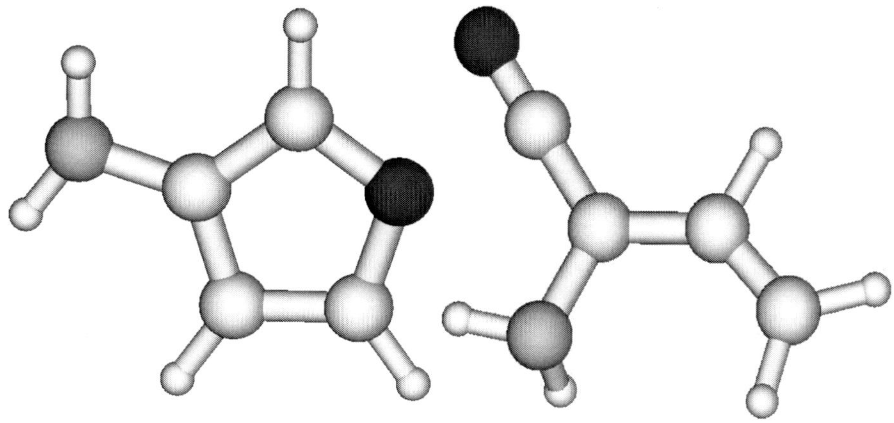

Figure 11. View of the $C_4H_5ON^+$ fragment with the amine-derived furan (left) and linear (right) structure.

Note that the measured appearance energy of the m=101 a.m.u. fragment ($C_4H_7NO_2^+$) is 7.5±0.1 eV and this value is less than the calculated HOMO binding energy in [21]. The experimental ionization energy of the parent molecule is about 8.6 eV [5, 21, 35]. Hence, the formation of the positive $C_4H_7NO_2^+$ ion proceeds from the vibrationally excited initial neutral molecule with the detachment of the negative CH_3SH^- fragment. It should be mentioned that the geometrical structure of the $C_4H_7NO_2^+$ cation could be different due to the different –C–H bond rupture, i.e. migration of different H atoms. All possible isomers of this cation were studied theoretically by us. The structure of the cation obtained shows that its formation is due to the H13 hydrogen atom migration to the sulfur atom and the rupture of the S2–C3 bond with no change in the initial structure of molecule. At higher energies, beginning from 11.0±0.1 eV, the same processes lead to the appearance of the CH_4S^+ fragment. This means that at electron energies below the ionization threshold the positive charge is localized on the $C_4H_7NO_2$ fragment, whereas, at higher energies, this occurs on the sulfur-containing fragment.

The peak at m=131 a.m.u. in the mass-spectrum of the methionine molecule is assigned to the $C_5H_9NOS^+$ ion and corresponds to elimination of the fragment(s) with m=18 a.m.u. This dissociation channel dominates at the photon energy of 8.43 eV. The above ionized and neutral fragments may be produced according to the following schemes:

$$C_5H_{11}NO_2S + e \rightarrow \begin{cases} C_5H_9NOS^- + (OH{+}H)^+ + e & (22) \\ C_5H_9NOS^0 + (OH{+}H)^+ + 2e; & (23) \\ C_5H_9NOS^+ + (OH{+}H)^- + e & (24) \\ C_5H_9NOS^+ + (OH{+}H)^0 + 2e & (25) \end{cases}$$

Comparison of the experimental and calculated data shows that formation of the $C_5H_9NOS^+$ ion (m=131 a.m.u.) at the parent molecule dissociation is not accompanied by detachment of two fragments (HO + H) (see Table 10). The energy release required to produce the neutral water molecule lowers the calculated appearance energy for the $C_5H_9NOS^+$ ion by 6.39 eV. Therefore, the appearance of the peak at m=131 a.m.u. in the methionine mass spectrum is related to the neutral water molecule elimination.

Table 10. Calculated appearance energies (eV) for the C_5H_9NOS fragment produced from the neutral methionine molecule.

C_5H_9NOS (m=131 a.m.u.) charge	(HO + H) fragment charge	Methionine molecule geometry not changed*	Methionine molecule geometry changed**
-1	1	18.96	13.52
0	1	21	13.87
1	-1	15.21	13.35
1	0	18.68	10.81

*,** – see explanatory notes in Table 4.

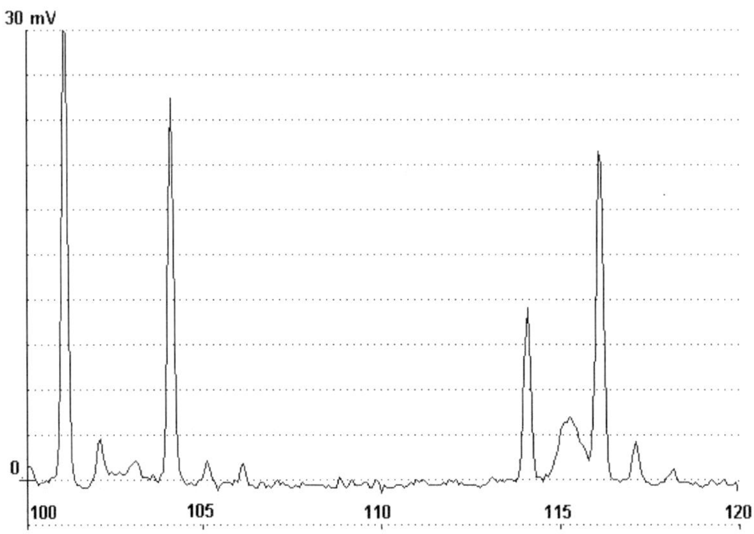

Figure 12. Large-mass area of the methionine molecule mass spectrum measured at 70 eV electron energy.

The large-mass area of the methionine molecule mass spectrum (Fig. 12) reveals a diffuse peak at about m^{*}~115.1 a.m.u. that corresponds to the 149→131 transition i.e. the elimination of the neutral fragment with m=18 a.m.u. Thus, the dissociation channel of the parent methionine molecule related to the neutral water molecule detachment was experimentally confirmed by us.

The methionine molecule mass-spectrum also reveals a peak at m=35 a.m.u. and its identification as belonging to the SH_3^+ fragment is at first sight doubtful. Indeed, formation of this fragment requires not only the break of two skeleton α-bonds with respect to the sulfur atom but also displacement of three hydrogen atoms. However, the experimental results prove that the peak at m=35 a.m.u. as that of the SH_3^+ one because it is accompanied by a satellite peak with m=37 a.m.u., the intensity of which corresponds to the natural ^{34}S isotope contribution.

Possibility to form the SH_3^+ cation when the C atom is removed from the CH_3S fragment is not confirmed, e.g., by the data on the methylthiole (CH_4S) and dimethyldisulfide (CH_3SSCH_3) mass spectrometry [5]. Both these compounds under electron impact produce the CH_3S^+ ion, though no SH_3^+ ion was observed in the relevant mass spectra. Our theoretical results exhibit the 6.27 – 6.44 eV energy required to remove the C atom from the CSH_3^+ cation. On the other hand, the mass spectra of dialkyl sulfides (say, CH_3SCH or $C_2H_5SC_2H_5$) reveal the $CH_3SCH_2^+$ and SH_3^+ ions [5]. When the dialkyl sulfides have alkyl branching that blocks off the $CH_3SCH_2^+$ ion formation, the SH_3^+ ion is not produced as well. This indirectly confirms the cascade character of the fragmentation resulting in the SH_3^+ ion appearance. It should be mentioned that the possible mechanism of the SH_3^+ cation formation from $CH_3SCH_2^+$ via the cyclic transition state is described in [34]. Therefore, the analysis of the mass spectra of the low-molecular dialkyl sulfides allows us to conclude that the sulfonium ion in the methionine molecule mass-spectrum arises due to the cascade mechanism of the $CH_3SCH_2^+$ ion secondary dissociation.

The semi-logarithmic scale presentation of the initial area of mass-spectrum (see Fig. 10) allowed the specific features (not discussed in the up-to-date papers) to be found, i.e. two weak peaks related to the doubly-charged $C_2H_3ON^{2+}$ (m/z=28.5) and $C_4H_3O^{2+}$ or $C_4H_5N^{2+}$ (m/z=33.5) ions. We seem to be the first to observe the above doubly-charged ions in the amine acid mass-spectra.

Obviously, the peak at m/z=28.5 is assigned to the doubly-charged $C_2H_3ON^{2+}$ ion. This conclusion is made based on the basis of the results obtained in this chapter. In the methionine molecule mass-spectrum, the singly-charged $C_2H_3ON^+$ ion with m=57 a.m.u. is present (Fig. 5). This indicates that the parent molecule fragmentation according to the 'amine' type could lead to formation of both singly- and doubly-charged ions with the 57 a.m.u. mass. Most probably, formation of the $C_2H_3ON^{2+}$ ion under electron impact occurs as a result of dissociation of the skeleton C4–C5 bond of the doubly-charged parent molecule ion accompanied by the water molecule elimination. Taking into account that the hydroxyl group and the hydrogen atom required for the water molecule formation are in the opposite parts of a parent molecule after the C4–C5 bond rupture, one may conclude that H_2O is eliminated simultaneously or slightly before the skeleton bond dissociates.

The results of optimization indicate that in the case of $[OH + H]^0$ and $[OH + H]^-$ a water molecule should be produced. Unfortunately, weak intensity of the $C_2H_3ON^{2+}$ ion peak did not allow us to measure the dissociative ionization function for this ion and to determine its appearance energy experimentally.

We suggest the following probable scheme of the doubly-charged $C_2H_3ON^{2+}$ ion formation:

$$C_5H_{11}NO_2S + e \rightarrow C_5H_{11}NO_2S^{2+} + 3e \rightarrow C_2H_3ON^{2+} + (H_2O + C_3H_6S)^0 + 3e \quad (26)$$

As stated above, Fig. 10 clearly illustrates an additional peak related to another doubly-charged ion at *m/z*=33.5. No singly-charged ion with the *m*=67 a.m.u. mass is present in the methionine molecule mass-spectrum. The gross formula of the doubly-charged ion at *m/z*=33.5 may be $C_4H_3O^{2+}$ or $C_4H_5N^{2+}$. Obviously, formation of the doubly-charged $C_4H_3O^{2+}$ ion occurs after elimination of the stable neutral fragments H_2O, NH_2, CH_4S from the doubly-charged parent molecule ion (pathway (27)). In the case of the $C_4H_5N^{2+}$ ion the parent doubly-charged molecule eliminates H_2O, CH_4S and the oxygen atom (pathway (28)). To check the possibility of the stepwise formation of this doubly-charged ion we have analyzed the areas of the mass spectra where the possible intermediate ions may arise. The doubly-charged $(M–H_2O)^{2+}$, $(M–NH_2)^{2+}$ and $(M–CH_4S)^{2+}$ ions corresponding to the sequential detachment of neutral fragments from the parent ion are not observed in the methionine molecule mass-spectrum, what allows us to exclude the cascade process during formation of the ion with *m/z*=33.5. Different possibilities of the doubly-charged ion formation were theoretically investigated. The results obtained allow us to conclude that the possible channels of formation of the doubly-charged ions at *m/z*=33.5 are as follows:

$$C_5H_{11}NO_2S + e \rightarrow C_5H_{11}NO_2S^{2+} + 3e \rightarrow C_4H_3O^{2+} + NH_2^0 + H_2O^0 + CH_4S^0 + 3e \quad (27)$$

$$\searrow C_4H_5N^{2+} + O^0 + H_2O^0 + CH_4S^0 + 3e \quad (28)$$

We provided calculations concerning the stability of the doubly-ionized fragments taking into account the structure of ions produced (Table 11).

Table 11. Stability of the doubly-ionized fragments with *m/z*=33.5.

Fragment	Internal energy, a.u.	Energy of formation, eV	Energy of formation per atom, eV
$C_4H_5N^{2+}$	-209.164	30.237	3.022
$C_4H_3O^{2+}$ (O=C-C-CH-CH$_2$)	-228.328	18.894	2.362
$C_4H_3O^{2+}$ (O=C-CH-CH-CH)	-228.340	18.59	2.324
$C_4H_3O^{2+}$ (Cyclic form)	-228.240	16.17	2.022

Our calculations of the internal energy for cyclic and different linear structures of the $C_4H_3O^{2+}$ ion allow us to conclude that the linear structure is more stable than the cyclic one. But, according to our calculations, among the doubly-charged ions with *m/z*=33.5 the most stable is the $C_4H_5N^{2+}$ one. The structure of this ion obtained after optimization, i.e. when the equilibrium geometry is achieved, is presented in Fig. 13.

Thus, after the loss of two electrons from the first two highest occupied molecular orbitals (i.e. non-bonding *n*-orbitals of the sulfur atom and the nitrogen atom) by the parent methionine molecule, formation of the doubly-charged $C_2H_3ON^{2+}$ (*m/z*=28.5) and the $C_4H_5N^{2+}$ and/or $C_4H_3O^{2+}$ (*m/z*=33.5) ions becomes possible. However, among the last ones the presence of the $C_4H_5N^{2+}$ ion is more probable due to its stability.

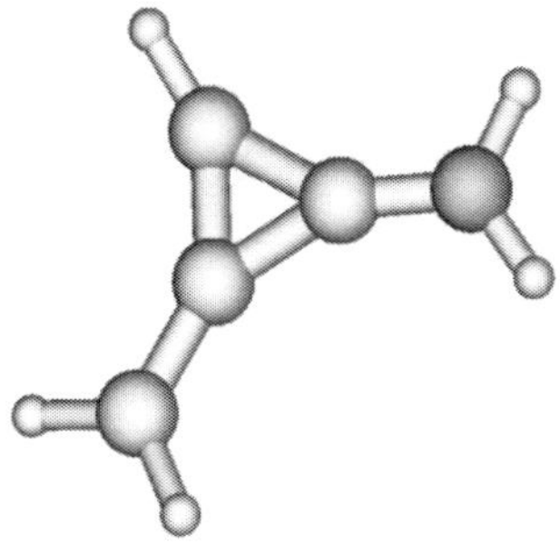

Figure 13. Structure of the $C_4H_5N^{2+}$ ion obtained after optimization.

Table 12. Appearance energies for the methionine fragments.

Fragment mass, a.m.u.	Ion assignment	Experimental appearance energy, eV	NIST database data [5]	Calculated data: Appearance energy, eV	Calculated data: Other products
28	CH_2N^+	11.4 ± 0.1	–	11.31**	$(C_3H_7S + H + CHO_2)^0$
30	CH_4N^+	11.0 ± 0.1	–	9.08**	$(COO + C_3H_7S)^0$
41	$C_3H_5^+$	13.3 ± 0.1	–	9.74**	$(NH_2 + CH_3S + CHO_2)^0$
	$C_2H_3N^+$			14.41**	$(CHO_2 + C_2H_5S + 2H)^-$
42	$C_3H_6^+$	12.9 ± 0.1	–	–	–
	$C_2H_4N^+$			9.37** 16.98**	$(C_2H_5S + H + CHO_2)^-$ $(C_2H_5S + H + CHO_2)^0$
43	$C_3H_7^+$	13.2 ± 0.1	–	–	–
	$C_2H_5N^+$			13.58* 7.46**	$(C_2H_5S + CHO_2)^-$
44	$C_3H_8^+$	12.9 ± 0.1		–	–
	$C_2H_6N^+$			8.28**	$(C_2H_5S + CO_2)^0$
	CS^+		–	16.76**	$(CH_3 + 2H + C_3H_6NO_2)^0$
45	CHO_2^+	13.5 ± 0.1	–	11.76**	$C_4H_{10}NS^0$
46	$CH_2O_2^+$	13.1 ± 0.1	–	13.22**	$C_4H_9NS^{0,-1}$
47	CH_3S^+	13.0 ± 0.1	13.0 ± 0.2	11.82**	$(C_4O_2NH_8)^0$
48	CH_4S^+	11.0 ± 0.1	–	10.57**	$C_4H_7NO_2^0$
55	$C_3H_5N^+$	11.8 ± 0.1	–	11.86**	$(CH_3S + 2H + CHO_2)^-$
56	$C_3H_6N^+$	11.2 ± 0.1	–	10.04*	$(CH_3S + H + CO_2H)^0$
57	$C_3H_7N^+$	12.7 ± 0.1	–	12.51**	$(CH_3S + CO_2H)^-$
61	$C_2H_5S^+$	12.4 ± 0.1	12.43 ± 0.1	10.71**	$C_3H_6NO_2^0$
74	$C_3H_6S^+$	–		12.26**	$C_2H_5NO_2^0$
	$C_2H_4NO_2^+$		–	9.79**	$C_3H_7S^0$
75	$C_3H_7S^+$	–		8.07**	$C_2H_4NO_2^-$
	$C_2H_5NO_2^+$		–	10.82**	$C_3H_6S^-$
88	$C_3H_6NO_2^+$	–	–	8.72**	$C_2H_5S^0$
101	$C_4H_7NO_2^+$	7.5 ± 0.1	–	9.26**	$(CH_3S + H)^0$
104	$C_4H_{10}NS^+$	9.7 ± 0.1	9.68 ± 0.15	8.7**	CHO_2^0
114	$C_5H_6OS^+$	9.3 ± 0.1	–	10.50**	$(NH + OH + 3H)^-$
116	$C_5H_8OS^+$	8.5 ± 0.1		14.93**	$(NH_2 + OH)^0$
	$C_5H_{10}NS^+$		–	14.37**	$(OH + O)^-$
131	$C_5H_9NOS^+$	–	–	10.81**	$(OH + H)^0$
149	$C_5H_{11}NO_2S^+$	8.6 ± 0.1	8.63 ± 0.10	8.09**	–

*,** – see explanatory notes in Table 4.

As mentioned above, we have determined experimentally the appearance energies for the methionine molecule fragments and complemented them with the relevant calculations (see, e.g. [35]). Table 12 shows our experimental and calculated by us appearance energies for the methionine fragments as compared with the data of the other papers.

We would like to note that in some cases the appearance energies of some fragments were not calculated because there was not possibility to obtain these fragments from the methionine molecule in its equilibrium point and was impossible to predict how the geometry of the methionine molecule could change when the above fragments are formed.

On the other hand, we have calculated the appearance energies of some fragments despite the fact that the experimental results for different reasons were not obtained. So, the intensity of the peak with *m*=85 a.m.u. is very weak, therefore it was not investigated experimentally although the results of theoretical investigations indicate that this fragment is $C_4H_5O_2^+$ and the energy of its appearance is equal to 11.17 eV.

However, despite all the modeling and calculations performed, the calculated appearance energy for the *m*=116 a.m.u. fragment is too high as compared to the value measured. We predict that this fragment, as well as those with *m*=42, 43, 44 a.m.u., are formed from methionine not in its equilibrium state. Other states possible for the formation are difficult to be predicted or investigated.

Acknowledgments

The authors would like to thank Prof. J. Bojarski and Prof. M. Cegla from Jagellonian University (Krakow, Poland) and V. Patasiene for technical support, Prof. P.D. Burrow from Lincoln University (Nebraska, USA) and Prof. A. Imre for their kind assistance in preparing and carrying out the present experiments as well as for the fruitful discussions. Special thanks go to the InSpire and NGI.LT projects for the resources and support of theoretical calculations provided as well as to the COST MP0802 activity.

References

[1] S. Cristoni and L.R. Bernardi, *Mass.Spectr.Rev.* 22, 369 (2003).

[2] B.D. Michael and P.A. O'Neill, *Science* 287, 1603 (2000).

[3] S. Denifl, B. Sonnweber, G. Hanel, P. Scheier and T.D. Märk, *Int. J. Mass Spectrom.* 238, 47 (2004).

[4] *The Molecular Origins of Life,* Cambridge University Press, Cambridge (1998).

[5] National Institute of Standards, Standard Reference Database: *Chemistry Webbook* http://webbook.nist gov.

[6] MI 1201 *Mass Spectrometer. Instruction Manual*, SELMI Corporation, (2002).

[7] NuDAQ. ACL-8112 Series Enhenced Multi-Functions Data Acquisition Cards.

[8] ICP DAS, Data Acquisition Board PISO–DA2, User Manuals. *http://holit.ua.*

[9] ICP DAS, Data Acquisition Board PISO–813, User Manuals. *http://holit.ua.*

[10] E.I. Selifonova, R.K. Chernova and O.E. Koblova, *Izvestia Saratov. Univer. Ser. Chem. Biol. Ecol.* 8, 23 (2008).

[11] G. Hanel, B. Gstir, T. Fiegele, F. Hagelberg, K. Becker, P. Scheier and A. Snegursky, *J.Chem. Phys.* 116, 2456 (2002).

[12] A.N. Zavilopulo and A.V. Snegursky, *Techn.Phys.Lett.* 28, 913 (2002).

[13] R.S. Freund, R.C. Wetzel, R.J. Shul and T.R. Hayes, *Phys.Rev. A.* 41, 3575 (1990).

[14] A.D. Becke, *J. Chem. Phys.* 98, 5648 (1993).

[15] R.A. Kendall, T.H. Dunning Jr and R.J.J. Harrison, *Chem. Phys.* 96, 6796 (1992).

[16] J. Tamuliene, L.G. Romanova, V.S. Vukstich and A.V. Snegursky, *Chem. Phys.* 404, 36 (2012).

[17] M.W. Schmidt, K.K. Baldrige, J.A. Boatz, S.T. Elbert, M.S. Gordon, J.H. Jensen, S. Koseki, N. Matsunaga, K.A. Nguyen, S.J. Su, T.L. Windus, M. Dupuis and J.A. Montgomery, *J. Comp. Chem.* 14, 1347 (1993).

[18] Gaussian 03, Revision C.02, Gaussian, Inc., Wallingford CT, (2004).

[19] A.T. Lebedev, *Mass-Spectrometry in Organic Chemistry*, BINOM, Moscow (2003).

[20] U. Salzner and R.J. Baer, *Chem. Phys.* 131, 231101 (2009).

[21] O. Plekan, V. Feyer, R. Richter, M. Coreno, M. Simone, K.C. Prince and V. Carravetta, *J.Phys.Chem. A* 111, 10998 (2007).

[22] L.S. Cederbaum, *J. Phys. B.* 8, 290 (1975).

[23] D. Dehareng and G. Dive, *Int. J. Mol. Sci.* 5, 301 (2004).

[24] H. Jürgen, *Gross Mass Spectrometry. A Textbook.* 2nd Edition. Springer-Verlag, Berlin-Heidelberg (2011).

[25] O. Plekan, V. Feyer, R. Richter, M. Coreno, M. de Simone and K.C. Prince. *Chem. Phys.* 334, 53 (2007).

[26] V.V. Afrosimov, A.A. Basalaev, E.A. Berezovskaya, M.N. Panov, O.V. Smirnov and A.V. Tulub, *Techn. Phys.* 51, 1553 (2006).

[27] R. Johnstone, *Mass-Spectrometry for Organic Chemists.* Cambridge University Press, Cambridge (1972).

[28] R. S. Mulliken, *J. Chem. Phys.* 23, 1833 (1955).

[29] J. Tamuliene, L.G. Romanova, V.S. Vukstich and A.V. Snegursky, *Nucl. Instr. Meth. B.* 279, 128 (2012).

[30] A.F. Lago, L.H. Coutinho, R.R.T. Marinho, A. Naves de Brito and G.G.B. de Souza *Chem. Phys.* 307, 9 (2004).

[31] H.-W. Jochims, M. Schwell and J.-L. Chotin, *Chem. Phys.* 298, 279 (2004).

[32] K. Biemann, J. Seibl and F.J. Gapp, *Am. Chem. Soc.* 89, 2719 (1967).

[33] G.A. Junk and H.J.J. Svec, *J. Am. Chem. Soc.* 85, 839 (1963).

[34] K.V. Rajesh, C. Majumder, O.D. Jayakumar, S.K. Kulshreshtaha and J.P. Mittal *Proc. Indian Acad. Sci. (Chem. Sci.)* 113, 129 (2001).

[35] J. Tamuliene, L.G. Romanova, V.S. Vukstich and A.V. Snegursky, *Chem. Phys.* 404, 74 (2012).

In: Methionine
Editor: Alexander Snegursky
ISBN: 978-1-62618-246-2

Chapter VIII

Protective Roles of Transsulfuration against Methionine Toxicity

Shotaro Kamata and Isao Ishii
Department of Biochemistry,
Keio University Graduate School of Pharmaceutical Sciences, Tokyo, Japan

Abstract

Methionine is an essential amino acid in mammals, but is the most toxic among the constituent amino acids of proteins. Excessive methionine intake causes oxidative/nitrosative injuries in the liver, hepatic/splenic hypertrophy, altered erythrocyte morphology, and the resultant splenic hemosiderosis in rats; however, the pathophysiology of methionine-induced liver toxicity remains to be elucidated. Methionine can be metabolized into cysteine by methionine cycle and transsulfuration, and cysteine can be further metabolized into cytoprotective metabolites such as glutathione, taurine, and hydrogen sulfide (H_2S). We recently found that a methionine-rich diet induces acute lethal hepatitis in transsulfuration-defective cystathionine γ-lyase-deficient mice. Molecular mechanisms by which methionine excess induces hepatic dysfunction and physiological roles of transsulfuration in its detoxification (metabolism) are herein reviewed.

1. Methionine Toxicity *In Vivo*

Methionine is an essential amino acid in mammals; however, it is clearly the most toxic among the constituent amino acids of proteins, whether expressed in terms of percentage of the diet, molality, or an increase over the requirement (0.5–0.6% of the diet) [1]. When the L-methionine level in voluntary food intake is maintained at four to six times the estimated requirement, food intake and growth are suppressed [2–8]. Its D-enantiomer is also effective [2,9,10] probably because of its rapid conversion to L-enantiomer *in vivo*; almost all intravenously administrated D-methionine is converted to L-methionine in rats [11]. Relative oral bioavailability values of D-methionine, with those of L-methionine set at 100%, are 90%, 90%, 75%, 100%, 100%, and 30%, for chick, rat, mouse, pig, dog, and human, respectively [12].

Excessive methionine intake causes oxidative/nitrosative injuries [6,13–16], hepatic/splenic hypertrophy [2,4], and splenic hemosiderosis [4,7] (summarized in Table 1). Some metabolic change or adaptation may occur because supplementation of serine or glycine could restore growth after a few days of feeding on a high-methionine diet in rats [3] and patients with innate hypermethioninemia by methionine adenosyltransferase (MAT) I/III deficiency are asymptomatic [17].

Methionine is normally metabolized in the liver to cysteine via the methionine cycle and transsulfuration (Figure 1) [21]. Serine is a substrate of the transsulfuration enzyme, cystathionine β-synthase (CBS), and glycine is a precursor of serine; therefore, supplementation of serine or glycine may lead to effective clearance of homocysteine (and thus, methionine) by using CBS activity.

Table 1. Toxicity of methionine and its related compounds.

Supplement	Species	Dose	Duration	Symptoms	Ref
L- or D-Met	rat	3.0% in diet	1–9 mo	growth↓, hepatic/splenic hypertrophy	[2]
L-Met	rat	3.0% in diet	2 wk	growth↓ (reversed by Gly & Ser)	[3]
L-Met	rat	3.0% in diet	2 wk	growth↓, splenic hypertrophy, splenic iron↑	[4]
L-Met	rat	3.0% in diet	1 wk	hepatic catalase/Cyt c oxidase↓	[13]
L-Met	rat	1.5% in diet	2 wk	food intake↓, growth↓	[5]
L-Met	rat	1.6% in diet	2 wk	growth↓, hepatic iron/ferritin/TBARS↑	[6]
L-Met	rat	2.4% in diet	2 wk	food intake↓, growth↓, splenic iron↑	[7]
L-Met	mouse (male) hepatocytes	30 mM	3–5 hr	cell viability↓, GSH↓	[18]
L-Met	rabbit	2.0% in diet	2 mo	oxidative stress (serum, heart, aorta)↑	[14]
L-Met	rat	2.0% in diet	6 mo	oxidative stress (serum, heart, aorta)↑	[15]
L-Met	mouse	2.2% in diet	2 wk	growth↓, hepatitis, splenic atrophy	[8]
DL-Met	rat	2.0% in diet	2 yr	growth↓, vascular aging	[9]
DL-Met	rat	2.0% in diet	7 wk	hepatic iron/TBARS↑	[16]
DL-Met	rat	2.0% in diet	6 wk	growth↓	[10]
N-acetyl-L-Met	rat	3.0% in diet	2 wk	growth↓, splenic hypertrophy, splenic iron↑	[4]
MTP	rat	2.57% in diet	2 wk	food intake↓, growth↓, splenic hypertrophy,altered erythrocyte morphology	[19]
MTP	mouse (male) hepatocytes	0.3 mM	2–5 hr	cell viability↓, GSH↓	[18]
L-Met-SO	mouse (male) hepatocytes	20-30 mM	3–5 hr	cell viability↓, GSH↓	[20]

mo, months; wk, weeks; hr, hours; yr, years; TBARS, thiobarbituric acid-reactive substrates.

On the other hand, excessive methionine can be converted via transamination to 2-keto-4-methylthiobutyrate (KMTB), 3-methylthiopropionate (MTP), and methanethiol (CH_3SH; methyl mercaptane) (Figure 2), all of which are extremely toxic [1,19,22]. Increased amounts of methionine transamination products are present in plasma and urine of patients with severe hypermethioninemia due to either MAT I/III deficiency or CBS deficiency [17]. Poisonous ammonia (NH_3) is also generated from glutamate, a transamination product from a reaction between methionine and α-ketoglutarate. There is a significant association between the severity grades of hepatic encephalopathy and serum concentrations of methionine, KMTB, NH_3, or CH_3SH-mixed disulfides [22]. Excess CH_3SH/NH_3 causes hepatic encephalopathy/coma in experimental animals [23].

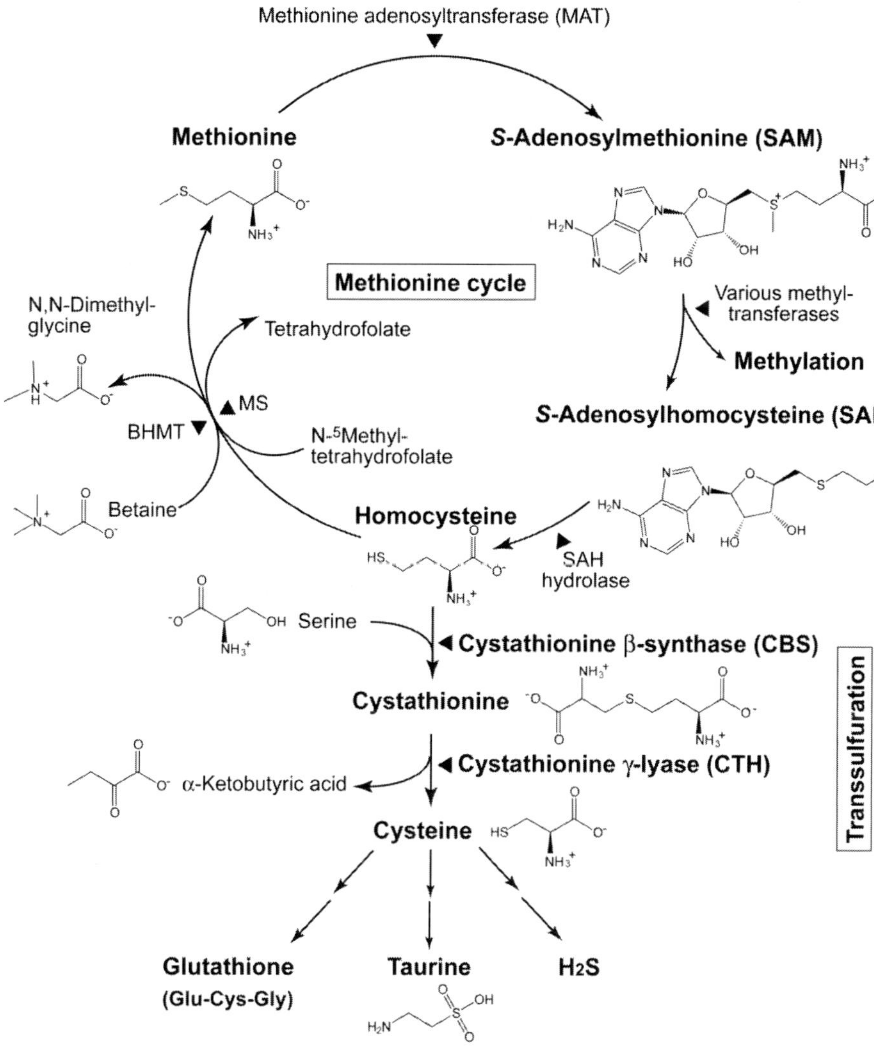

Figure 1. Methionine cycle and transsulfuration. Methionine is converted by MAT to SAM, which is irreversibly converted to SAH by donating its methyl moiety to various methyltransferases. SAH hydrolase regulates the bidirectional reaction, leading to the synthesis of homocysteine from SAH or vice versa. Homocysteine is converted to methionine (called re-methylation) or to cysteine via transsulfuration pathway. In the transsulfuration, CBS catalyzes a condensation reaction between homocysteine and serine to form cystathionine, and CTH catalyzes its hydrolysis to form cysteine. Biosynthesized cysteine is utilized to form glutathione, taurine, and H_2S. BHMT, betaine-homocysteine S-methyltransferase; MS, methionine synthase.

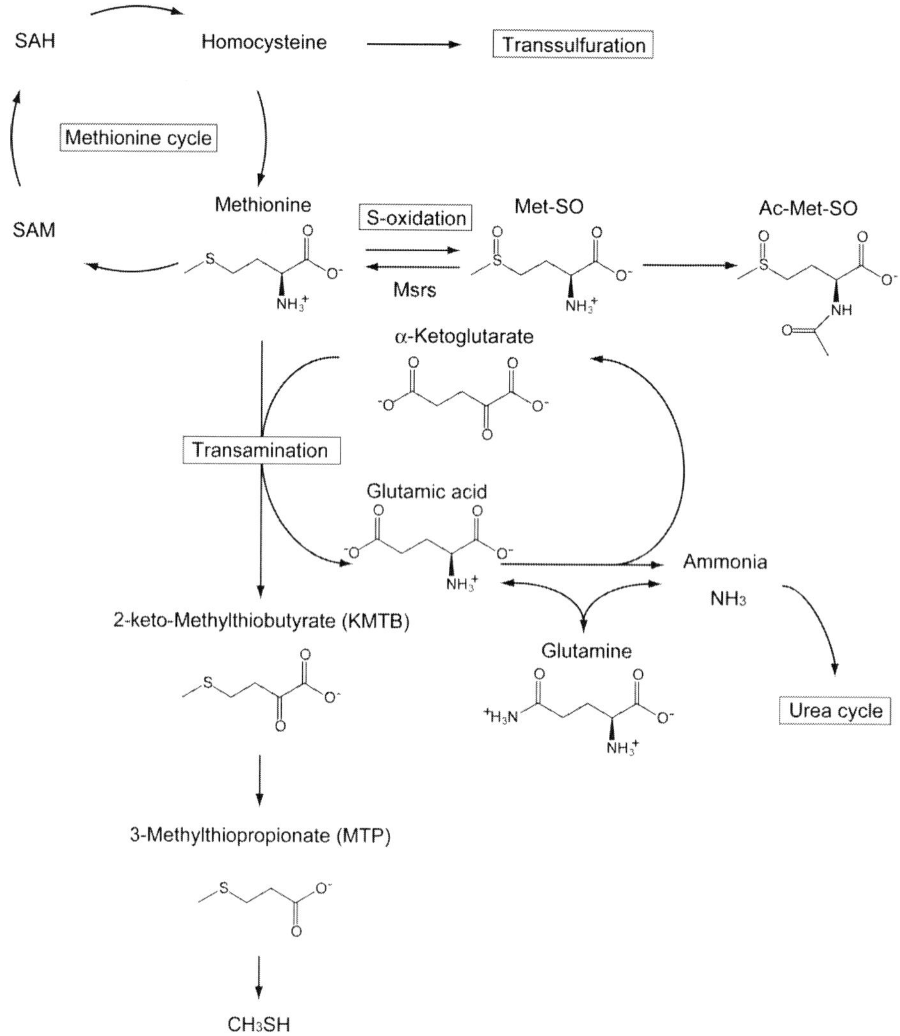

Figure 2. Methionine transamination and oxidation pathway. Methionine can be converted via transamination to KMTB and then MTP or via oxidation to Met-SO. Some of Met-SO is metabolized to its acetylated form, Ac-Met-SO. Anti-oxidant enzymes, methionine sulfoxide reductases (Msrs), mediate reduction of free or protein-based Met-SO to methionine.

Moreover, either free or protein-based methionine is susceptible to oxidation by reactive oxygen species (ROS) at its sulfur atom under physiological conditions; the major product is cytotoxic methionine sulfoxide (Met-SO) (Figure 2) [8,20,24]. There are two diastereomers, Met-*S*-SO and Met-*R*-SO. The oxidation of methionine may be reversed by three methionine sulfoxide reductase (Msr) families (MsrA, MsrB, and fRMsr) with the help of thioredoxin and glutaredoxin systems. MsrA and MsrB are specific for the reduction of free and protein-based Met-*S*-SO and Met-*R*-SO, respectively, whereas fRMsr (free methionine-*R*-sulfoxide reductase) is highly specific for free Met-*R*-SO and does not reduce protein-based sulfoxides [24]. The principal function of Msrs is to repair oxidatively damaged proteins by reversing the oxidation of methionine residues, and for this reason, Msrs are called as anti-oxidant enzymes. Recent studies using MsrA-deficient mice revealed that MsrA is involved in the protection of mouse kidney from ischemia-reperfusion injury [25] and mouse neurons from Amyloid β-protein insults [26] as well as in the reduced accumulation of protein-carbonyls in various mouse tissues [27].

While several toxic methionine metabolites are found, *S*-adenosylmethionine (SAM; a MAT product) is sold as a nutritional supplement to improve liver function and depression under the marketing name SAM-e [28]. Its precise working mechanisms are yet unknown, but its supplementation is considered to up-regulate cysteine biosynthesis. Biosynthesized cysteine is utilized to generate major antioxidants, glutathione, (hypo)taurine, and H_2S, which may counteract with oxidative toxicity by excessive methionine. The pathophysiology of methionine-induced toxicity remains to be elucidated.

2. Protective Roles of Transsulfuration

To date, six genetic mouse models with targeted deletion of methionine cycle/transsulfuration enzymes (except multiple SAM-dependent methyltransferases) have been reported: mice lacking *S*-adenosylhomocysteine (SAH) hydrolase [29], methionine synthase [30], MAT I/III [31], CBS [32], BHMT [33], and cystathionine γ-lyase (CTH) [34,35]; the first two of these are found to be embryonically lethal. MAT I/III-deficient mice are apparently normal but predisposed to liver injury [31] and spontaneously develop liver tumors [36], whereas CBS-deficient ($Cbs^{-/-}$) mice (a model of homocystinuria; OMIM ID: 236200) suffer from severe hepatic dysfunction/steatosis and most die before 4 weeks of age [32,37]; importantly, both mice display hypermethioninemia [31,37]. BHMT-deficient mice are susceptible to fatty liver and hepatocellular carcinoma [33]. To reveal the roles of transsulfuration against methionine toxicity, we examined the pathogenesis upon excessive methionine intake utilizing transsulfuration-defective CTH-deficient ($Cth^{-/-}$) mice, a genetic model of cystathioninuria (OMIM ID: 219500) that we generated recently [35]. Homozygous ($Cth^{-/-}$) mice were obtained from heterozygous ($Cth^{+/-}$) intercrosses at the expected frequency, and developed with no apparent abnormalities [35]. Unlike $Cbs^{-/-}$ mice, $Cth^{-/-}$ mice were free of hypermethioninemia and hepatic dysfunction [35]. However, $Cth^{-/-}$ mice required dietary cyst(e)ine (or its protein source) to protect against acute lethal myopathy because cysteine is an essential amino acid for $Cth^{-/-}$ mice, clearly demonstrating that CTH is an essential enzyme for transsulfuration [35].

Wild-type ($Cth^{+/+}$), $Cth^{+/-}$, and $Cth^{-/-}$ females were fed 1×Met (0.44% methionine), 3×Met (1.32%), or 6×Met (2.64%) diet after weaning at 3 weeks of age. All mice developed comparably on 1×Met with no obvious phenotypic abnormality [8]. Meanwhile, only $Cth^{-/-}$ mice showed retarded growth on 3×Met and daily weight loss on 6×Met until dying within 2 weeks. After one week with the diets, serum levels of alanine/aspartate aminotransferases, alkaline phosphatase, and total bilirubin were markedly elevated in 6×Met-fed $Cth^{-/-}$ mice compared with those in 6×Met-fed wild-type mice or 3×Met-fed $Cth^{-/-}$ mice [8]. Histological examination revealed inflammatory cell infiltration around some central veins in 3×Met-fed $Cth^{-/-}$ mice, although much more severe infiltration and hepatocyte loss were observed around all central veins in 6×Met-fed $Cth^{-/-}$ mice, and hepatic antioxidant capacity was severely impaired in 6×Met-fed $Cth^{-/-}$ mice [8]. Another apparent feature of 6×Met-fed $Cth^{-/-}$ mice was splenic atrophy; tissue to body weight ratios were elevated in the brain, heart, and kidney, and unchanged in the liver, while those in the spleen were significantly decreased after one week on 6×Met, although hemosiderosis was not observed by Prussian blue staining [8].

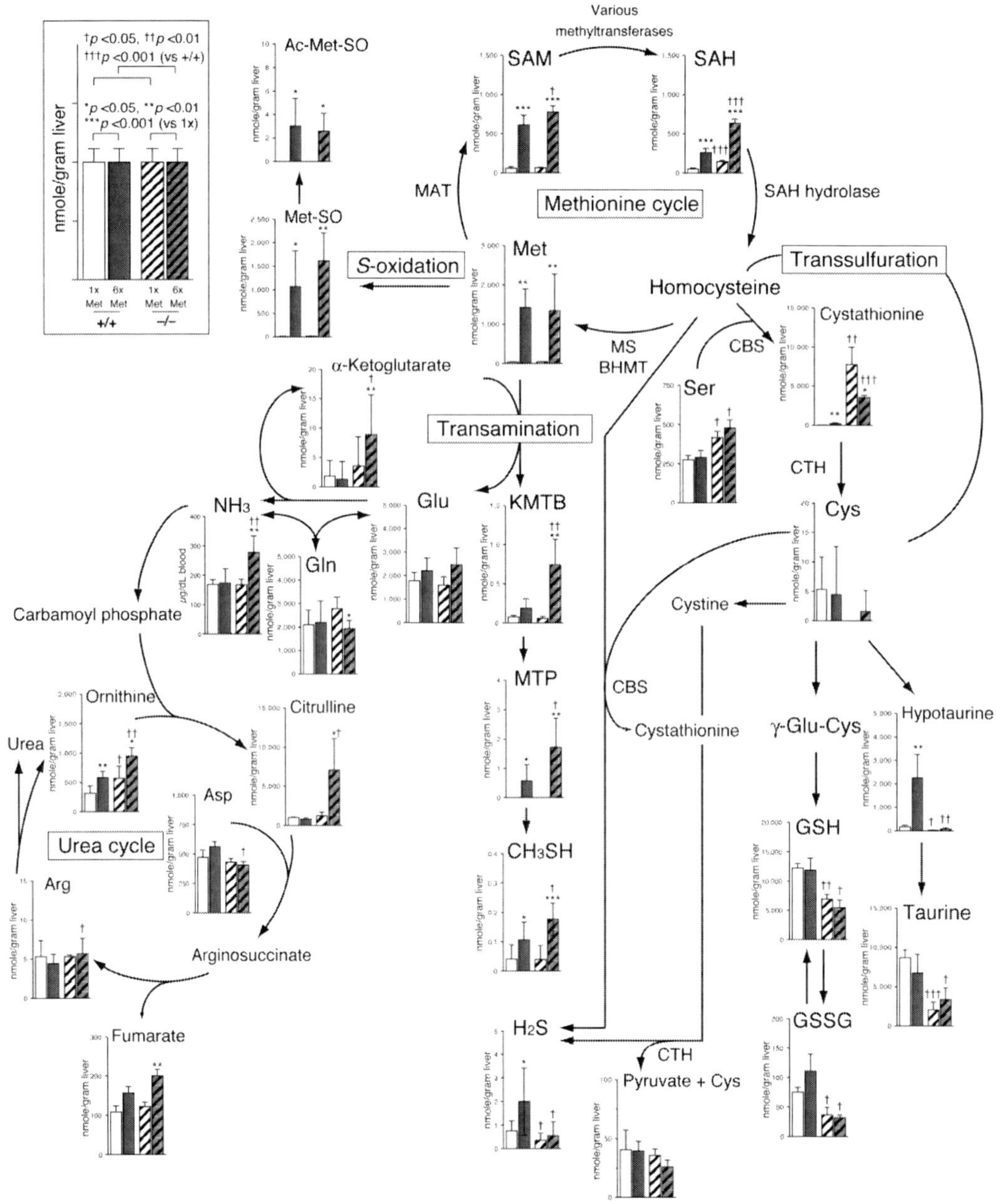

Figure 3. Metabolome analysis of methionine derivatives in the livers of wild-type and *Cth*$^{-/-}$ mice that were fed 1×Met or 6×Met for 1 week after 3 weeks of age. Small molecule metabolites in liver homogenates were quantified using CE-MS, LC-MS, gas chromatography, CoulArray, and Dri-Chem. Blood levels of NH_3 were measured using Dri-Chem. Values are the mean ± SD (n = 5–11); $*p < 0.05$, $**p < 0.01$, and $***p < 0.001$ *versus* 1×Met-fed animals; and $^{\dagger}p < 0.05$, $^{\dagger\dagger}p < 0.01$, and $^{\dagger\dagger\dagger}p < 0.001$ *versus* wild-type animals. (Adapted from [8] with permission from Elsevier).

One of the common features of homocystinuria patients and *Cbs*$^{-/-}$ mice is hyper-accumulation of methionine in blood (~2 mM in humans [17] and ~2.5 mM in mice [37]); this feature was not observed in *Cth*$^{-/-}$ mice [35]. Examination of serum amino acid levels after one week on 6×Met revealed hypermethioninemia (1.44 mM) as well as hyperhomocysteinemia (256 μM total homocysteine) in *Cth*$^{-/-}$ mice, which resembles the biochemical features of *Cbs*$^{-/-}$ mice at 2 weeks of age (2.6 mM and 209 μM, respectively) that suffer from severe hepatic steatosis [37]. However, oil red O staining confirmed that 6×Met-fed *Cth*$^{-/-}$ mice were free of hepatic steatosis, unlike 2-week-old *Cbs*$^{-/-}$ mice [8].

Autopsy revealed that all dead *Cth*$^{-/-}$ mice had intestinal bleeding and some dying *Cth*$^{-/-}$ mice displayed gastric bleeding with no sign of ulcers; such bleeding was not observed in other organs, including the brain, lung, thymus, spleen, kidney, and testis. Bleeding time and platelet numbers were unaltered between wild-type and *Cth*$^{-/-}$ mice after one week on 6×Met, whereas prothrombin time and activated partial thromboplastin time were prolonged and fibrinogen levels were decreased in 6×Met-fed *Cth*$^{-/-}$ mice [8]. Therefore, the life-threatening bleeding is likely caused by coagulation factor deficiency from hepatic dysfunction rather than disseminated intravascular coagulation (DIC).

Metabolome analysis assisted by capillary electrophoresis-mass spectrometry (CE-MS), liquid chromatography-mass spectrometry (LC-MS), sulfur-detective gas chromatography, coulometric electrochemical detection (CoulArray), and automated clinical chemistry analyzer (Dri-Chem) has allowed us to identify changes in methionine metabolites under pathologic conditions systematically.

One-week feeding on 6×Met induced a significant increase in hepatic levels of methionine, SAM, SAH, Met-SO, Ac-Met-SO (a methionine sulfoxide metabolite), KTMB, MTP, and CH_3SH in both wild-type and *Cth*$^{-/-}$ mice; among them, the levels of SAM, SAH, KMTB, MTP, and CH_3SH were significantly higher in *Cth*$^{-/-}$ mice. In contrast, blood levels of NH_3 and hepatic levels of α-ketoglutarate and some urea cycle metabolites (citrulline and fumarate) were significantly elevated in 6×Met-fed *Cth*$^{-/-}$ mice although not in 6×Met-fed wild-type mice. One-week feeding on 6×Met al.so induced a significant decrease in its downstream anti-oxidative metabolites, such as reduced glutathione (GSH)/oxidized glutathione (GSSG), taurine/hypotaurine, and H_2S, confirming that transsulfuration was severely impaired [8]. In the absence of transsulfuration, excessive methionine was primarily metabolized to Met-SO via *S*-oxidation or to KTMB/glutamate via transamination.

Conclusion

Transsulfuration-defective *Cth*$^{-/-}$ mice displayed acute lethal hepatitis in cases of excessive methionine intake, probably from oxidative hepatotoxicity by the combination of increased toxic transamination metabolites/homocysteine/Met-SO levels and the decreased overall antioxidant capacity (Figure 4). Therefore, the transsulfuration pathway is essential to protect against methionine toxicity *in vivo*.

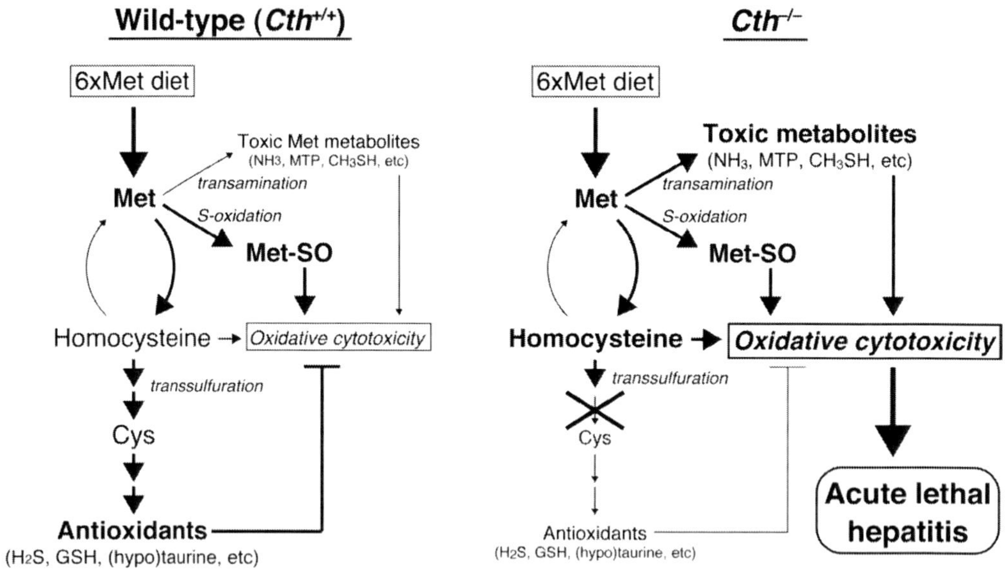

Figure 4. Schematic representation of methionine metabolism in the livers of 6×Met diet-fed wild-type or *Cth*$^{-/-}$ mice. (Adapted from [8] with permission from Elsevier).

Acknowledgments

This work was supported in parts by Grants-in-Aid for Scientific Research (No. 22590292), Innovative Areas (No. 23117716), and Program for Strategic Research Foundation at Private Universities (2011–2015) from the Ministry of Education, Culture, Sports, Science and Technology of Japan.

References

[1] Benevenga, N. J., and Steele, R. D. (1984). Adverse effects of excessive consumption of amino acids. *Annu Rev Nutr*, 4, 157-181.

[2] Stekol, J. A., and Szaran, J. (1962). Pathological effects of excessive methionine in the diet of growing rats. *J Nutr*, 77, 81-90.

[3] Benevenga, N. J., and Harper, A. E. (1967). Alleviation of methionine and homocystine toxicity in the rat. *J Nutr*, 93, 44-52.

[4] Rotruck, J. T., and Boggs, R. W. (1977). Effects of excess dietary l-methionine and N-acetyl-l-methionine on growing rats. *J Nutr*, 107, 357-362.

[5] Regina, M., Korhonen, V. P., Smith, T. K., Alakuijala, L., and Eloranta, T. O. (1993). Methionine toxicity in the rat in relation to hepatic accumulation of S-adenosylmethionine: prevention by dietary stimulation of the hepatic transsulfuration pathway. *Arch Biochem Biophys*, 300, 598-607.

[6] Mori, N., and Hirayama, K. (2000). Long-Term Consumption of a Methionine-Supplemented Diet Increases Iron and Lipid Peroxide Levels in Rat Liver. *J. Nutr.*, 130, 2349-2355.

[7] Toue, S., Kodama, R., Amao, M., Kawamata, Y., Kimura, T., and Sakai, R. (2006). Screening of toxicity biomarkers for methionine excess in rats. *J Nutr*, 136, 1716S-1721S.

[8] Yamada, H., Akahoshi, N., Kamata, S., Hagiya, Y., Hishiki, T., Nagahata, Y., Matsuura, T., Takano, N., Mori, M., Ishizaki, Y., Izumi, T., Kumagai, Y., Kasahara, T., Suematsu, M., and Ishii, I. (2012). Methionine excess in diet induces acute lethal hepatitis in mice lacking cystathionine gamma-lyase, an animal model of cystathioninuria. *Free Radic Biol Med*, 52, 1716-1726.

[9] Fau, D., Peret, J., and Hadjiisky, P. (1988). Effects of ingestion of high protein or excess methionine diets by rats for two years. *J Nutr*, 118, 128-133.

[10] Amaral, C. L., Bueno Rde, B., Burim, R. V., Queiroz, R. H., Bianchi Mde, L., and Antunes, L. M. (2011). The effects of dietary supplementation of methionine on genomic stability and p53 gene promoter methylation in rats. *Mutat Res*, 722, 78-83.

[11] Hasegawa, H., Shinohara, Y., Akahane, K., and Hashimoto, T. (2005). Direct detection and evaluation of conversion of D-methionine into L-methionine in rats by stable isotope methodology. *J Nutr*, 135, 2001-2005.

[12] Baker, D. H. (2006). Comparative species utilization and toxicity of sulfur amino acids. *J Nutr*, 136, 1670S-1675S.

[13] Finkelstein, A., and Benevenga, N. J. (1986). The effect of methanethiol and methionine toxicity on the activities of cytochrome c oxidase and enzymes involved in protection from peroxidative damage. *J Nutr*, 116, 204-215.

[14] Yalcinkaya Demirsoz, S., Depboylu, B., Dogru-Abbasoglu, S., Unlucerci, Y., and Uysal, M. (2009). Effects of high methionine diet on oxidative stress in serum, apo-B containing lipoproteins, heart, and aorta in rabbits. *Ann Clin Lab Sci*, 39, 386-391.

[15] Yalcinkaya, S., Unlucerci, Y., Giris, M., Olgac, V., Dogru-Abbasoglu, S., and Uysal, M. (2009). Oxidative and nitrosative stress and apoptosis in the liver of rats fed on high methionine diet: protective effect of taurine. *Nutrition*, 25, 436-444.

[16] Lynch, S. M., and Strain, J. J. (1989). Increased hepatic lipid peroxidation with methionine toxicity in the rat. *Free Radic Res Commun*, 5, 221-226.

[17] Mudd, S. H., Levy, H. L., and Kraus, J. P. Disorders of transsulfuration. In: Scriver, C. R., Beaudet, A. L., Sly, W. S., and Valle, D., editors. *The metabolic and molecular basis of inherited disease* (8th ed). New York: McGraw-Hill; 2001; 2007-2056.

[18] Dever, J. T., and Elfarra, A. A. (2008). L-Methionine toxicity in freshly-isolated mouse hepatocytes is gender-dependent and mediated in part by transamination. *J Pharmacol Exp Ther*, 326, 809-817.

[19] Steele, R. D., Barber, T. A., Lalich, J., and Benevenga, N. J. (1979). Effects of dietary 3-methylthiopropionate on metabolism, growth and hematopoiesis in the rat. *J Nutr*, 109, 1739-1751.

[20] Dever, J. T., and Elfarra, A. A. (2008). L-methionine-dl-sulfoxide metabolism and toxicity in freshly isolated mouse hepatocytes: gender differences and inhibition with aminooxyacetic acid. *Drug Metab Dispos*, 36, 2252-2260.

[21] Stipanuk, M. H. (2004). Sulfur amino acid metabolism: pathways for production and removal of homocysteine and cysteine. *Annu Rev Nutr*, 24, 539-577.

[22] Blom, H. J., Ferenci, P., Grimm, G., Yap, S. H., and Tangerman, A. (1991). The role of methanethiol in the pathogenesis of hepatic encephalopathy. *Hepatology*, 13, 445-454.

[23] Zieve, L. (1987). Pathogenesis of hepatic encephalopathy. *Metab Brain Dis*, 2, 147-165.

[24] Lee, B. C., and Gladyshev, V. N. (2011). The biological significance of methionine sulfoxide stereochemistry. *Free Radic Biol Med*, 50, 221-227.

[25] Kim, J. I., Choi, S. H., Jung, K. J., Lee, E., Kim, H. Y., and Park, K. M. (2012). Protective Role of Methionine Sulfoxide Reductase A Against Ischemia/Reperfusion Injury in Mouse Kidney and Its Involvement in the Regulation of Trans-Sulfuration Pathway. *Antioxid Redox Signal*, in press.

[26] Moskovitz, J., Maiti, P., Lopes, D. H., Oien, D. B., Attar, A., Liu, T., Mittal, S., Hayes, J., and Bitan, G. (2011). Induction of methionine-sulfoxide reductases protects neurons from amyloid beta-protein insults in vitro and in vivo. *Biochemistry*, 50, 10687-10697.

[27] Moskovitz, J., and Oien, D. B. (2010). Protein carbonyl and the methionine sulfoxide reductase system. *Antioxid Redox Signal*, 12, 405-415.

[28] Martinez-Chantar, M. L., Garcia-Trevijano, E. R., Latasa, M. U., Perez-Mato, I., Sanchez del Pino, M. M., Corrales, F. J., Avila, M. A., and Mato, J. M. (2002). Importance of a deficiency in S-adenosyl-L-methionine synthesis in the pathogenesis of liver injury. *Am J Clin Nutr*, 76, 1177S-1182S.

[29] Miller, M. W., Duhl, D. M., Winkes, B. M., Arredondo-Vega, F., Saxon, P. J., Wolff, G. L., Epstein, C. J., Hershfield, M. S., and Barsh, G. S. (1994). The mouse lethal nonagouti (a(x)) mutation deletes the S-adenosylhomocysteine hydrolase (Ahcy) gene. *EMBO J*, 13, 1806-1816.

[30] Swanson, D. A., Liu, M. L., Baker, P. J., Garrett, L., Stitzel, M., Wu, J., Harris, M., Banerjee, R., Shane, B., and Brody, L. C. (2001). Targeted disruption of the methionine synthase gene in mice. *Mol Cell Biol*, 21, 1058-1065.

[31] Lu, S. C., Alvarez, L., Huang, Z. Z., Chen, L., An, W., Corrales, F. J., Avila, M. A., Kanel, G., and Mato, J. M. (2001). Methionine adenosyltransferase 1A knockout mice are predisposed to liver injury and exhibit increased expression of genes involved in proliferation. *Proc Natl Acad Sci USA*, 98, 5560-5565.

[32] Watanabe, M., Osada, J., Aratani, Y., Kluckman, K., Reddick, R., Malinow, M. R., and Maeda, N. (1995). Mice deficient in cystathionine beta-synthase: animal models for mild and severe homocyst(e)inemia. *Proc Natl Acad Sci USA*, 92, 1585-1589.

[33] Teng, Y. W., Mehedint, M. G., Garrow, T. A., and Zeisel, S. H. (2011). Deletion of betaine-homocysteine S-methyltransferase in mice perturbs choline and 1-carbon metabolism, resulting in fatty liver and hepatocellular carcinomas. *J Biol Chem*, 286, 36258-36267.

[34] Yang, G., Wu, L., Jiang, B., Yang, W., Qi, J., Cao, K., Meng, Q., Mustafa, A. K., Mu, W., Zhang, S., Snyder, S. H., and Wang, R. (2008). H_2S as a physiologic vasorelaxant: hypertension in mice with deletion of cystathionine gamma-lyase. *Science*, 322, 587-590.

[35] Ishii, I., Akahoshi, N., Yamada, H., Nakano, S., Izumi, T., and Suematsu, M. (2010). Cystathionine gamma-lyase-deficient mice require dietary cysteine to protect against acute lethal myopathy and oxidative injury. *J Biol Chem*, 285, 26358–26368.

[36] Martinez-Chantar, M. L., Corrales, F. J., Martinez-Cruz, L. A., Garcia-Trevijano, E. R., Huang, Z. Z., Chen, L., Kanel, G., Avila, M. A., Mato, J. M., and Lu, S. C. (2002).

Spontaneous oxidative stress and liver tumors in mice lacking methionine adenosyltransferase 1A. *FASEB J*, 16, 1292-1294.

[37] Akahoshi, N., Kobayashi, C., Ishizaki, Y., Izumi, T., Himi, T., Suematsu, M., and Ishii, I. (2008). Genetic background conversion ameliorates semi-lethality and permits behavioral analyses in cystathionine beta-synthase-deficient mice, an animal model for hyperhomocysteinemia. *Hum Mol Genet*, 17, 1994-2005.

In: Methionine
Editor: Alexander Snegursky
ISBN: 978-1-62618-246-2

Chapter IX

Methionine, Homocysteine and Cysteine, and Antiepileptic Drugs in Epilepsy

***Jolanta Dorszewska*[1*] *and Wojciech Kozubski*[2]**
[1]Laboratory of Neurobiology, Department of Neurology,
[2]Chair and Department of Neurology,
Poznan University of Medical Sciences, Poznan, Poland

Abstract

The last decade has seen many studies on hyperhomocysteinemia (HHcy) in epileptic patients treated with antiepileptic drugs (AEDs). It has been shown that HHcy occurs in 10-40% of the epileptic population and that AEDs pharmacotherapy has a fundamental effect on the levels of homocysteine (Hcy). The literature indicates that carbamazepine (CBZ) and valproic acid (VPA) therapy in epileptics leads to an increase of Hcy levels. It has also been indicated that the levels of cysteine (Cys) in epileptic patients treated with AEDs change, which may result in disturbed transformation of Cys to glutathione (GSH), an important factor in the maintenance of redox homeostasis in the body.

The study included 63 epileptic patients and 38-61 controls. The levels of Hcy, Met, Cys and GSH were determined by high-performance liquid chromatography (HPLC) with electrochemical detection.

The present study demonstrates that AEDs treatment in epileptics leads to an increase in Hcy ($p<0.001$) and GSH ($p<0.01$), a decrease in Cys ($p<0.001$), and relatively unchanged methionine (Met) concentrations, with disturbed levels of Met:Hcy ($p<0.01$) and Cys:Hcy ($p<0.001$) ratios. Carbamazepine monotherapy ($p<0.05$) and AEDs polytherapy ($p<0.001$) and long-term therapy ($p<0.001$) compared to controls appear to be related to disturbances in both Met:Hcy, and Cys:Hcy ratios. Additionally, it was found that after VPA or CBZ monotherapy and long-term therapy of the two AEDs together lead to a reduction of the level in Cys (VPA, $p<0.05$; long-term therapy, $p<0.01$

* Corresponding author: Jolanta Dorszewska, MDs, PhD, Laboratory of Neurobiology, Department of Neurology, Poznan University of Medical Sciences, 49 Przybyszewskiego St., 60-355 Poznan, Poland. Phone: ++ 48 61 86 91 439; Fax: ++ 48 61 86 91 697; Email: dorszewskaj@yahoo.com

compared to controls) and increased the level of GSH (CBZ, $p<0.001$; long-term therapy, $p<0.01$ compared to controls). In patients with epilepsy before treatment with AEDs, there were slightly changed Hcy and Cys levels but significantly reduced levels of Met ($p<0.05$) and statistically insignificantly increased levels of GSH.

It seems that AEDs pharmacotherapy, rather than epilepsy, leads to increases in Hcy levels, especially in patients treated in polytherapy and long-term. In patients with epilepsy before AEDs therapy, the remethylation process of Hcy appears to be disrupted and after AEDs treatment, especially after VPA therapy, there seems to be a disruption in the transsulfuration process of this thiol. Our study indicated that oxidative stress (measured as levels of GSH) was increased in patients with epilepsy both before and after treatment with AEDs (in particular CBZ and long-term therapy), but it appears to occur via different mechanisms. It also seems that, in epileptic patients treated with AEDs, only therapy with B-vitamins and folates can effectively help regulate the levels of the vascular disease factor Hcy and the metabolism of biothiols (Hcy, Met, Cys, GSH).

Introduction

The last decade has seen many studies on hyperhomocysteinemia (HHcy) in epileptic patients treated with antiepileptic drugs (AEDs). It has been shown that HHcy occurs in 10-40% of the epileptic population [1,2], and that AEDs pharmacotherapy has a fundamental effect on the levels of homocysteine (Hcy) in both children [3-5] and adults [6-10].

The literature indicates that carbamazepine (CBZ) therapy in epileptics leads to an increase of Hcy levels [4,5,8,11]. Treatment with valproic acid (VPA), however, has variable effects; in some cases, the Hcy level is decreased [12], in others it is increased [3-5,8], and in yet others VPA has no effect on Hcy levels [7,13,14]. Lamotrigine (LTG) treatment of epileptic patients has not been shown to lead to an increase in Hcy [12,15].

Furthermore, it has been demonstrated that a specific predisposition to generate circulating Hcy in patients with gene polymorphisms in both the cystathionine beta-synthase (*CBS*) and methylenetetrahydrofolate reductase (*MTHFR*) C677T [2,7] and A1298C [16,17] genes is an important factor in the levels of HHcy in epileptics treated with AEDs. The study by Mudd et al.. [18] shows that 20% of the heterozygote CBS patients have epileptic seizures when circulating Hcy increases from 50 to 200 μM. It has also been shown that, in women with generalized idiopathic epilepsy, the frequency of the *MTHFR*, TT (C677T) homozygote is increased, with this polymorphism also possibly increasing the risk of becoming an epileptic [19]. The literature indicates that Hcy formation in epileptics treated with AEDs may also involve 5,10-methylenetetrahydrofolate reductase (MTR). Studies on animal models have shown that phenytoin administration can both decrease [20] and increase [21] MTR activity.

It is also known that the three-functional enzyme methylenetetrahydrofolate dehydrogenase/ methynyltetrahydrofolate cyclohydrolase/ formyltetrahydrofolate synthetaze (MTHFD1) is involved in the regulation of circulating Hcy. It has been shown that *MTHFD1* homozygotes experience an increase in Hcy levels in heart disease and dysgenesis of the neural tube dependent on folate (FA) levels [22], while *MTHFD1* GA (G1958A) heterozygotes with Alzheimer's or Parkinson's disease have variable cysteine (Cys) to Hcy ratios (Cys:Hcy) [23].

Moreover, it has been demonstrated that both heterozygote and homozygote individuals with the *MTHFR* (C677T) polymorphism have a significantly increased level of Hcy and decreased levels of FA, especially in polytherapy with AEDs [7,13,24]. However, the study by Sniezawska et al.. [10] shows greater increases in Hcy concentration during AEDs treatment of epileptics with the *MTHFR* CT (C677T) and *MTHFD1* GG (G1958A) genotypes.

Additionally, it has been shown that the MTHFT, MTR, and MTHFD1 enzymes take part in converting Hcy to methionine (Met), and changes in the activity of these enzymes can have an effect on the effectiveness of the process of remethylation from Hcy to Met, as well as the levels of these compounds [21]. Ono et al.. [25] demonstrated that the level of Met is decreased in epileptics treated with AEDs. However, the study by Sniezawska et al.. [10] shows greater decreases in Met concentration during AEDs treatment of epileptics with the *MTHFR* CC and CT (C677T), *MTHFD1* GG and GA (G1958A), and *MTR* AA and AG (A2756G) genotypes.

In the body, Hcy is a point of intersection of two main metabolic pathways: transsulfuration and remethylation. Under physiological conditions, approximately 50% of Hcy is catabolized by transsulfuration and undergoes transformation to Cys and then to glutathione (GSH). The remaining 50% of Hcy undergoes remethylation to Met. Methionine is supplied with food and its transformation to Hcy involves several steps. At the first step, Met is transformed to SAM (S-adenosylmethionine) and then to SAH (S-adenosylhomocysteine), and hydrolyzed to Hcy (Figure 1). Agnati et al.. [26] showed that Hcy may pass the blood/brain barrier and that the level of plasma Hcy corresponds to the Hcy concentration in the brain. Elevated levels of Hcy in the central nervous system (CNS) may damage vascular endothelium, deteriorate the functionality of the blood/brain barrier, lead to disturbed production of nitrogen oxide, and to neurotoxic effects both in the senescent brain and in neurological diseases.

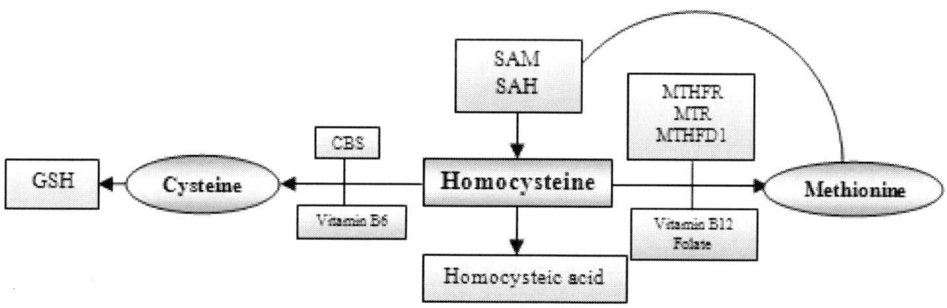

Figure 1. Synthesis and metabolic pathways of homocysteine, CBS- cystationine β-synthase, GSH- glutathione, MTHFR- 5,10-methylenetetrahydrofolate reductase, MTR- methionine synthase, MTHFD1- methylenetetrahydrofolate dehydrogenase/ methenyltetrahydrofolate cyclohydrolase/formyltetrahydrofolate synthetase, SAH- S-adenosylhomocysteine, SAM-S-adenosylmethionine.

Homocysteine or it oxidative product, homocysteic acid (Figure 1), are reported as producers of convulsions (Figure 2). In animal models, administration of Hcy leads to generalized seizures that can be prevented by N-methyl-D-aspartate (NMDA) or non-NMDA receptor antagonists [27,28]. The study by Folbergrova et al.. [29] demonstrates prevention of seizures induced by homocysteic acid in rats using a group II metabotropic glutamate receptor (mGluR) agonist. Simultaneously, in the rat cortex, hippocampus, and brain stem, a cyclic product of Hcy, thiolactone, significantly inhibits Na+/K+-ATPase activity, which may contribute at least in part to the understanding of the excitotoxic and convulsive properties of Hcy [30]. In humans, only in homocystinuric patients but not in patients with HHcy, seizures have been observed due to a CBS deficit [31]. The role of HHcy in the development of oxidative stress and DNA damage to the genes encoding proteins involved in epileptogenesis must also be considered.

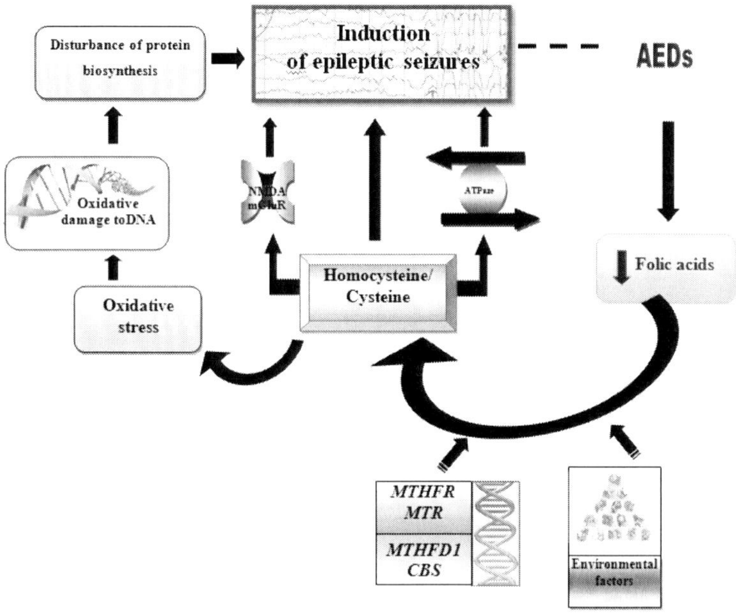

Figure 2. Homocysteine, cysteine and epileptic seizures, AEDs- antiepileptic drugs, CBS- cystationine β-synthase, MTHFR- 5,10-methylenetetrahydrofolate reductase, MTR- methionine synthase, MTHFD1- methylenetetrahydrofolate dehydrogenase/ methenyltetrahydrofolate cyclohydrolase/formyltetrahydrofolate synthetase, NMDA- N-methyl-D-aspartate receptor, mGluR- group II metabotropic glutamate receptor.

Biothiols and Antiepileptic Drugs in Epileptic Patients

There is much published literature demonstrating an increase in the plasma concentration of Hcy in epileptic patients treated with AEDs. These articles indicate that pharmacologic treatment in epileptic patients can lead to an increase in Hcy concentration that may be due to various factors, including diet, the type of pharmacologic agents used in treatment, the length of treatment, and genetic factors.

It seems that one of the significant factors affecting the generation of Hcy in epileptic patients treated with AEDs is the patient's diet, especially the presence of essential cofactors for the metabolism of Hcy (vitamins B6 and B12, FA). It also appears that the demonstrated lack of Hcy concentration increase after AEDs seen in studied American populations [12,32, 33], in contrast to European (Greece, [5]; Germany, [6]; Italy, [3]; Norway, [7]; Poland, [10,34]), Japanese [25], Korean [13], Turkish [4,15], and Egyptian [8] populations, is most likely due to the diet consumed by the majority of Americans, a diet rich in the cofactors necessary for remethylation of Hcy to Met (FA) [35].

It is also known that insufficiency of vitamin B12 and low FA levels may lead to a decreased efficiency in the remethylation process of Hcy, which results in a decrease of Met concentration. The studies by Ono et al.. [25] and Sniezawska et al.. [10] show decreased Met concentrations in patients with epilepsy after AEDs treatment. Additionally, the study by Badiou et al.. [11] in epileptic patients with HHcy shows increased Cys levels after CBZ treatment as compared to oxcarbazepine (OXCBZ) therapy. However, the study by Apeland et al.. [9] in patients with epilepsy treated with AEDs demonstrates an increased level of total Cys with perturbation of the redox status of biothiols. Similar changes have been observed in the aminothiol redox status in patients with cardiovascular disease, where an alteration in plasma redox was evidenced by the decrease in thiol/disulfide ratios of Cys, Hcy, and cysteinylglycine. In all analyzed groups in this study, Cys was directly correlated with Hcy but not with GSH or cysteinylglycine, which in turn were correlated each other. It seems that in patients with heart diseases the levels of plasma Cys were more linked to Hcy than to the metabolism of GSH [36].

There is little information in the literature on the correlation between Hcy, Cys and GSH in patients with epilepsy treated with AEDs.

Homocysteine and Methionine Levels and Antiepileptic Drugs

The purpose of this study was to analyse the levels of the sulphur-containing amino acids Hcy and Met in epileptic patients treated with VPA, CBZ, new generation (NG) AEDs and patients treated by polytherapy, as well as patients before AEDs treatment and control group.

Patients

The studies were conducted on 63 patients with idiopathic, cryptogenic epilepsy, including 28 women and 35 men, aged 18-65 years (mean age: 35.3±13.8 years). Among the epileptic patients, 55 of them, 24 women and 31 men aged 18-65 years (mean age: 35.8±13.4 years), were treated with variable AEDs. Eight patients, 4 women and 4 men aged 18-65 years (mean age: 31.7±17.2 years), were preparing to begin AEDs treatment.

Among the patients treated with AEDs, 67% of them were on monotherapy [VPA, 59%; CBZ, 27%; NG – mainly LTG and OXCBZ – 14%], and 33% were receiving polytherapy. In the polytherapy group, 61% were taking VPA and NG AEDs, 22% were taking VPA and CBZ, and 17% were taking CBZ and NG AEDs. In the overall group of patients receiving AEDs, 31% had been receiving AEDs for fewer than 5 years, while 69% had been receiving AEDs for over 5 years.

In the group of patients treated with AEDs, the mean levels of VPA and CBZ were within the reference values and averaged 57.1 μg/ml and 6.8 μg/ml, respectively.

The control group included 61 individuals, 41 women and 20 men, aged 22-67 years (mean age: 44.3±14.2 years).

The classification of epilepsy and/or epileptic syndrome suffered by each patient was verified according to the criteria and terminology recommended by the Commission on Classification and Terminology of The International League Against Epilepsy [37].

None of the epileptic patients and control subjects (epileptic patients and controls – Caucasian), were diagnosed with impaired liver or kidney function. None of the patients and controls had been receiving B group vitamin or FA supplementation, and testing was conducted a few hours before the patients took their AEDs dose.

None of the control subjects had verifiable symptoms of dementia or any other neurological disorders or smoking and drinking habits, and receiving B group vitamin or FA supplementation.

A Local Ethical Committee approved the study and the written consent of all patients or their caregivers was obtained.

Analysis of Hcy and Met Concentrations

Preparation of Samples

The analyzed plasma thiol compounds (Hcy, Fluka Germany; Met, Sigma, USA) were diluted with water at a 2:1 ratio and reduced using 1% TCEP [Tris-(2-carboxyethyl)-phosphin-hydrochloride; Applichem, Germany] at a 1:9 ratio. Subsequently, the sample was deproteinized using 1M $HClO_4$ (at a 2:1 ratio) and applied to the HPLC/EC system.

Determination of Hcy and Met Concentration

The samples were fed to the HPLC system (P580A; Dionex, Germany) coupled to an electrochemical detector (CoulArray 5600; ESA, USA). The analysis was performed in Termo Hypersil BDS C18 column (250mm x 4.6mm x 5μm) [Germany] in isocratic conditions, using the mobile phase of 0.15 M phosphate buffer, pH 2.9, supplemented with 12.5-17% acetonitrile for estimation of Hcy and Met [38].

The system was controlled and the data were collected and processed using Chromeleon software (Dionex, Germany).

Results

The levels of Hcy and Met were studied in the plasma of epileptic patients treated by AEDs in monotherapy (VPA, CBZ, NG AEDs), epileptics treated by polytherapy (VPA+NG AEDs, CBZ+NG AEDs, VPA+CBZ), epileptics before AEDs treatment, and the control group.

The study revealed that the epileptics being treated with various AEDs have a significant increase in their plasma concentration of Hcy (one-way ANOVA test, $p<0.05$ compared to controls) [Table 1]. It was also shown that an increase in Hcy concentration above the minimum required to diagnose mild HHcy (above 16 μM) occurred in 27% of the patients, while an increase above the normal range (above 15 μM) occurred in 36% of epileptics treated with AEDs. Furthermore, the concentration of Met was significantly decreased only in epileptics who had not yet commenced treatment with AEDs (one-way ANOVA test, $p<0.05$ compared to controls). In both patients treated with AEDs and awaiting AEDs treatment, a decreased Met:Hcy ratio was observed (one-way ANOVA test, for AEDs-treated patients, $p<0.01$ compared to controls).

Table 1. Concentrations of Hcy and Met in epileptic patients before AEDs treatment AEDs(-), after beginning AEDs treatment AEDs(+), and controls

Analyzed compound	Control group	Epileptic patients		p
		AEDs(-)	AEDs(+)	
Hcy [μM]	11.9 ± 4.0	12.9 ± 3.2	14.5 ± 6.7*	0.0383
Met [μM]	22.6 ± 6.0	15.7 ± 9.2*	19.7 ± 7.9	0.0180
Met:Hcy	2.1 ± 0.9	1.4 ± 0.6	1.5 ± 0.8**	0.0011

Results are presented as mean ± SD.
One-way ANOVA for independent variables was used.
Statistically significant differences at *$p<0.05$, **$p<0.01$ compared to controls.

Statistically significant differences in Spearman test:

Control group:

- Spearman R coefficient +0.6286, $p<0.0001$ between Met and Met:Hcy.

AED(-) patients:

- Spearman R coefficient +0.8108, $p=0.0269$ between Met and Met:Hcy.

AED(+) patients:

- Spearman R coefficient -0.3479, $p=0.0099$ between Hcy and Met:Hcy.
- Spearman R coefficient +0.6215, $p<0.0001$ between Met and Met:Hcy.

Taking into account the type of pharmacotherapy used in epileptic patients, it was observed (Table 2) that the concentration of Hcy was significantly increased only in the patients undergoing polytherapy (one-way ANOVA test, $p<0.01$ compared to controls) and there was an essentially unchanged Hcy concentration only in the patients taking NG AEDs. The concentration of Met was decreased the most in patients taking CBZ and essentially unchanged in those treated with VPA. Moreover, the patients treated with VPA did not demonstrate a decrease in the Met:Hcy ratio (one-way ANOVA test, for patients treated with CBZ, $p<0.05$ compared to controls; for patients treated by polytherapy, $p<0.001$ compared to controls).

Considering the length of time AEDs treatment was used, long-term (more than 5 years) treatment of epileptics with various AEDs (Table 3, Figure 3) lead to a significant increase of Hcy concentration (Mann-Whitney test, $p<0.001$ compared to patients treated for less than 5 years with AEDs). Furthermore, it was shown that long-term treatment with AEDs insignificantly decreased the levels of Met while significantly decreasing the Met:Hcy ratio (Mann-Whitney test, $p<0.001$ compared to short-term treatment with AEDs).

Table 2. Concentrations of Hcy and Met in epileptic patients treated with VPA, CBZ, NG AEDs, and polytherapy, AEDs(+) before starting AEDs(-) treatment, and controls

Analyzed compound	Control group	AEDs(-)	Epileptic patients				p
			VPA	CBZ	NG AEDs	Polytherapy	
Hcy [µM]	11.9 ± 4.0	12.9 ± 3.2	13.0 ± 5.0	14.2 ± 4.8	11.6 ± 4.1	17.2 ± 9.0**	0.0126
Met [µM]	22.6 ± 6.0	15.7 ± 9.2	22.9 ± 9.0	16.1 ± 5.4	17.6 ± 7.4	18.5 ± 6.5	0.0069
Met:Hcy	2.1 ± 0.9	1.4 ± 0.6	2.0 ± 0.9	1.2 ± 0.5*	1.6 ± 0.9	1.1 ± 0.3***	0.0001

Results are presented as mean ± SD.

One-way ANOVA for independent variables was used.

Statistically significant differences at $*p<0.05$ between patients treated with CBZ and controls, and at $**p<0.01$, $***p<0.001$ between patients treated with polytherapy and controls.

Table 3. Concentrations of Hcy and Met in epileptic patients, treated with various AEDs, depending on length of treatment

Analyzed compound	Epileptic patients treated with AEDs		P
	Less than 5 years	More than 5 years	
Hcy [μM]	10.7 ± 3.9	16.2 ± 7.0***	0.0001
Met [μM]	20.5 ± 8.5	19.4 ± 7.6	0.2603
Met:Hcy	2.0 ± 1.0	1.3 ± 0.6***	0.0010

Results are presented as mean ± SD.
Non-parametric Mann-Whitney test for independent variables was used.
Statistically significant differences at ***p<0.001 compared to patients treated with AEDs for less than 5 years.

It seems that generation of Hcy in epileptic patients is affected mainly not by the disease itself, but the pharmacologic treatment for the disease. This is demonstrated by studies conducted on a Greek population, in epileptic children aged 4.5 to 14 years treated by AEDs for 20 weeks [5], by studies conducted on epileptic adult patients treated with AEDs in mono- and polytherapy for over 30 days [6], and by a study conducted in patients awaiting treatment with AEDs. The last mentioned study was conducted by Verrotti et al.. [3] in children, and is similar to our present study that was conducted on patients aged 18 to 65 years.

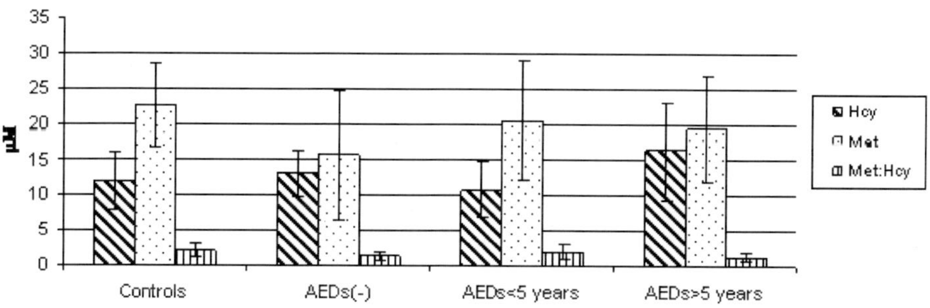

Results are presented as mean ± SD, one-way ANOVA test was used, for Hcy and Met:Hcy, p<0.001 between AEDs>5 years and controls, p<0.01 between AEDs<5 and AEDs>5 years.

Figure 3. Concentrations of Hcy and Met in epileptic patients with AEDs treatment less and more than 5 years, before AEDs(-) treatment, and controls.

The generation of Hcy in epileptic patients, as demonstrated by the literature, is related both to AEDs (such as phenytoin, phenobarbital, primidone, and CBZ) inducing microsomal liver enzymes and increasing cytochrome P450 enzyme activity, which leads to a decrease in FA serum concentration, as well as AEDs (such as VPA) that induce liver enzymes to a smaller degree and have a lesser effect on the concentration of FA in blood serum [39]. An increase in Hcy and accompanying decrease in FA concentration in epileptic patients after receiving CBZ was demonstrated in the studies by Verrotti et al.. [3], Karabiber et al.. [4] and Attilakos et al.. [5]. In epileptics taking VPA, an increase in Hcy concentration was demonstrated both accompanying a decrease in FA levels [3,4] and without FA level changes [5].

It is believed that a decrease in FA concentration during AEDs treatment is due to impaired intestinal absorption of FA, an increase in FA usage required for AEDs hydroxylation, the activation of liver enzymes leading to a final decrease in FA concentration, direct interactions between FA and AEDs caused by structural similarities between coenzymes for FA and AEDs, or a direct effect of AEDs on Hcy metabolism and kidney function [40].

However, there are also reports in the literature demonstrating a lack of AEDs effect both on inducing liver enzymes and on live enzyme activity changing the levels of circulating Hcy. Both in Sener et al.. [15] and in the present study, no significant increase in Hcy concentration was detected in epileptic patients treated with CBZ, while the articles by Apeland et al.. [9], Unal et al.. [14], and in the present study, there was no significant increase in Hcy after VPA treatment. The lack of significant increase in Hcy concentration in epileptic patients treated with CBZ and VPA may be due to these AEDs having only a small influence on the concentration of cofactors (the literature has shown definite effects of AEDs on the levels of FA and vitamin B12) essential to the remethylation of Hcy [4,5], and additionally in the case of VPA, the articles by Apeland et al.. [7], Gidal et al.. [12], and Attilakos et al.. [5] have shown that VPA increases the concentration of vitamin B12, which most probably prevents the accumulation of Hcy in the blood of epileptic patients.

It is also believed that NG AEDs such as LTG [12,15] and OXCBZ [11] do not change the concentrations of vitamin B12 and FA [12] in epileptic patients, nor do they increase the concentration of Hcy. The current study confirms these observations through the analysis of Hcy levels in epileptic patients receiving NG AEDs in monotherapy. Furthermore, many studies demonstrate that combining NG AEDs therapy, especially with VPA, can lead to increased formation of Hcy [6,8,17,25].

Another predisposing factor for increased formation of serum Hcy seems to be long-term (more than 5 years, [13]) therapy of epilepsy using AEDs. Increased levels of Hcy in patients receiving long-term treatment of CBZ, phenytoin, phenobarbital, and VPA and/or in polytherapy have also been demonstrated in the literature [8,25,41,42, present study]. The study by Hamed et al.. [8] demonstrates that long-term CBZ and VPA therapy take part in blood vessel damage via disturbed metabolism of Hcy and FA.

It is also known that insufficiency of the cofactors needed for the remethylation of Hcy (vitamin B12, FA) can lead to a decreased efficiency of this process, which results in a decrease of Met concentration [10]. A decrease in Met concentration in epileptic patients treated by mono- and polytherapy was demonstrated by Ono et al.. [25] as well as the present study, especially when using medication that decreases the levels of FA (CBZ). This study has also shown that a decrease in the Met:Hcy ratio accompanied the decrease of Met concentration. According to one of the newer theories [43,44], a decrease in the Met:Hcy ratio may be related to the transformation of Hcy to thiolactone in endothelium and mediate Hcy-induced atherosclerotic changes. Folates, which may increase Met and decrease Hcy levels, in vascular endothelial cells also inhibit conversion of Hcy to thiolactone by changing the relative levels accounting for the Hcy: Met ratio [44].

Homocysteine, Cysteine and Glutathione Levels and Antiepileptic Drugs

The purpose of this study was to analyse the levels of the sulphur-containing amino acids Cys and GSH in epileptic patients treated with VPA, CBZ, NG AEDs and patients treated by polytherapy, as well as patients before AEDs treatment and control group.

Patients

Epileptic patients: see point- Homocysteine and methionine levels and antiepileptic drugs.

The control group included 38 individuals, 28 women and 10 men, aged 22-67 years (mean age: 41.8±14.5 years).

Analysis of Hcy Concentrations

Analysis of Hcy levels: see point- Homocysteine and methionine levels and antiepileptic drugs.

Analysis of Cys and GSH Concentrations

Preparation of Samples

The analyzed plasma thiol compounds (Cys and GSH, Sigma, USA) were diluted with water at a 2:1 ratio and reduced using 1% TCEP [Tris-(2-carboxyethyl)-phosphin-hydrochloride; Applichem, Germany] at a 1:9 ratio. Subsequently, the sample was deproteinized using 1M $HClO_4$ (at a 2:1 ratio) and applied to the HPLC/EC system.

Determination of Cys and GSH Concentration

The samples were fed to the HPLC system (P580A; Dionex, Germany) coupled to an electrochemical detector (CoulArray 5600; ESA, USA). The analysis was performed in Termo Hypersil BDS C18 column (250mm x 4.6mm x 5μm) [Germany] in isocratic conditions, using the mobile phase of 0.15 M phosphate buffer, pH 2.9, supplemented with 12.5-17% acetonitrile for estimation of GSH and 0.15 M phosphate buffer, pH 2.8 with 10 % acetonitrile for estimation of Cys [33].

The system was controlled and the data were collected and processed using Chromeleon software (Dionex, Germany).

Results

The levels of Cys, GSH and Hcy were studied in the plasma of epileptic patients treated by AEDs in monotherapy (VPA, CBZ, NG AEDs), epileptics treated by polytherapy

(VPA+NG AEDs, CBZ+NG AEDs, VPA+CBZ), epileptics before AEDs treatment, and the control group.

The study revealed that epileptics being treated with various AEDs have a significant increase in their plasma concentration of Hcy (one-way ANOVA test, $p<0.001$ compared to controls) and GSH (one-way ANOVA test, $p<0.01$ compared to controls), and decreased Cys level (one-way ANOVA test, $p<0.001$ compared to controls) [Table 4]. Furthermore, the concentration of GSH was significantly increased only in epileptics who had been treated with AEDs. In both patients treated with AEDs and awaiting AEDs treatment, a decreased Cys:Hcy ratio was observed (one-way ANOVA test, for AEDs-treated patients, $p<0.001$ compared to controls).

Table 4. Concentrations of Hcy, Cys and GSH in epileptic patients before AEDs treatment AEDs(-), after beginning AEDs treatment AEDs(+), and controls

Analyzed compound	Control group	Epileptic patients		P
		AEDs(-)	AEDs(+)	
Hcy [μM]	10.0 ± 3.5	12.9 ± 3.2	14.5 ± 6.7***	0.0003
Cys [μM]	266.9 ± 56.3	264.8 ± 71.9	221.2 ± 56.6***	0.0009
Cys:Hcy	29.0 ± 13.3	23.2 ± 9.4	17.5 ± 7.6***	0.0011
GSH [μM]	888.3 ± 151.9	1071.7 ± 199.0	1042.3 ± 251.4**	0.0029

Results are presented as mean ± SD.
One-way ANOVA for independent variables was used.
Statistically significant differences at $^{**}p<0.01$, $^{***}p<0.001$ compared to controls.

Statistically significant differences in Spearman test:

Control group:

- Spearman R coefficient +0.6665, $p<0.0001$ between Cys and Cys:Hcy.

AED(+) patients:

- Spearman R coefficient -0.4295, $p=0.0012$ between Hcy and Cys:Hcy.
- Spearman R coefficient +0.6035, $p<0.0001$ between Cys and Cys:Hcy.

Taking into account the type of pharmacotherapy used in epileptic patients, it was shown (Table 5) that the concentration of Hcy was significantly increased only in the patients undergoing polytherapy (one-way ANOVA test, $p<0.001$ compared to controls). The concentration of Cys was decreased the most in patients taking NG AEDs and a significant in those treated with VPA (one-way ANOVA test, $p<0.05$ compared to controls). Moreover, on the patients treated only with NG AEDs did not demonstrated a significant decrease in the Cys:Hcy ratio (one-way ANOVA test, for patients treated with VPA, $p<0.01$; for patients treated with CBZ, $p<0.05$; for patients treated by polytherapy, $p<0.001$ compared to controls). Furthermore, the concentration of GSH was significantly increased only in epileptics who had treated with CBZ (one-way ANOVA test, $p<0.001$ compared to controls).

Considering the length of time AEDs treatment was used, long-term (more than 5 years) treatment of epileptics with various AEDs (Table 6, Figure 4) lead to a significant increase of Hcy concentration (Mann-Whitney test, $p<0.001$ compared to patients treated for less than 5 years with AEDs). Furthermore, it was shown that long-term treatment with AEDs did not

Table 5. Concentrations of Hcy, Cys and GSH in epileptic patients treated with VPA, CBZ, NG AEDs, and polytherapy, AEDs(+) before starting AEDs(-) treatment, and controls

Analyzed compound	Control group	AEDs(-)	Epileptic patients				p
			VPA	CBZ	NG AEDs	Polytherapy	
Hcy [μM]	10.0 ± 3.5	12.9 ± 3.2	13.0 ± 5.0	14.2 ± 4.8	11.6 ± 4.1	17.2 ± 9.0***	0.0002
Cys [μM]	266.9 ± 56.3	264.8 ± 71.9	220.7 ± 55.8*	210.1 ± 51.7	208.0 ± 49.9	232.2 ± 64.4	0.0099
Cys:Hcy	29.0 ± 13.3	23.2 ± 9.4	18.8 ± 7.3**	17.1 ± 8.8*	20.1 ± 8.4	15.4 ± 7.0***	0.0001
GSH [μM]	888.3 ± 151.9	1071.7 ±199.0	1014.1 ± 261.3	1232.3 ± 247.8***	930.0 ± 244.1	1000.2 ± 202.0	0.0005

Results are presented as mean ± SD.
One-way ANOVA for independent variables was used.
Statistically significant differences at *$p<0.05$ between patients treated with VPA, CBZ and controls, and at **$p<0.01$ between patients treated with VPA and controls, ***$p<0.001$ between patients treated with CBZ, polytherapy and controls.

change the levels of Cys but did significantly decrease the Cys:Hcy ratio (Mann-Whitney test, $p<0.01$ compared short-term treatment with AEDs), and insignificantly increased GSH concentrations.

Table 6. Concentrations of Hcy, Cys and GSH in epileptic patients, treated with various AEDs, depending on length of treatment

Analyzed compound	Epileptic patients treated with AEDs		P
	Less than 5 years	More than 5 years	
Hcy [μM]	10.7 ± 3.9	16.2 ± 7.0***	0.0001
Cys [μM]	221.0 ± 55.7	221.2 ± 57.9	0.4334
Cys:Hcy	22.3 ± 7.9	15.4 ± 6.4**	0.0030
GSH [μM]	1000.4 ± 235.4	1061.6 ± 259.2	0.3936

Results are presented as mean ± SD.
Non-parametric Mann-Whitney test for independent variables was used.
Statistically significant differences at $^{**}p<0.01$, $^{***}p<0.001$ compared to patients treated with AEDs for less than 5 years.

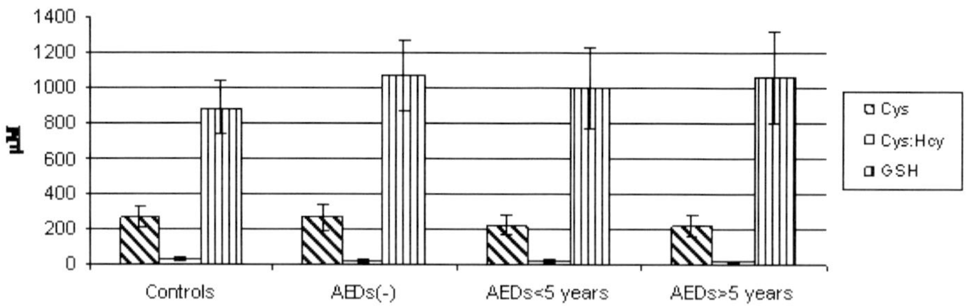

Results are presented as mean ± SD, one-way ANOVA test was used, for Cys, GSH, $p<0.01$ and Cys:Hcy $p<0.001$ between AEDs>5 years and controls; for Cys, $p<0.05$ between AEDs<5 years and controls.

Figure 4. Concentrations of Cys and GSH in epileptic patients with AEDs treatment less and more than 5 years, before AEDs(-) treatment, and controls.

Under physiological conditions, half of Hcy undergoes transsulfuration and provides Cys. The study by Bugaj et al.. [45] and the present study show that AEDs pharmacotherapy, rather than epilepsy, affects the transformation of Hcy to Cys and Hcy to Met. It also seems that AEDs therapy (especially CBZ and VPA therapies), which leads to decreased levels of Cys and slightly unchanged Met concentrations, is most likely associated with increased metabolism of Cys [45, present study]. However, the study by Badiou et al.. [11] demonstrated an increase in the level of Cys in epileptic patients with HHcy after more than 1 month of therapy with CBZ, which is probably associated with impaired remethylation of Hcy to Met. It appears that the length of AEDs treatment in epileptic patients may affect the level of biothiols. The study by Bochynska et al.. [34] shows a significantly increased level of Hcy in epileptic patients after 1 year of treatment with VPA and unchanged Hcy concentrations in patients with chronic epilepsy treated with VPA for at least 2 years.

The present study also showed a decreased Cys:Hcy ratio, especially in patients with epilepsy treated with AEDs in polytherapy and after long-term therapy. Changed levels of

Cys and Hcy may be associated with impaired saturation in protein and enzymes by the reaction of thiol-disulfide exchange and further modification of protein properties [46]. Additionally, the increase in Hcy and decrease in Cys levels after AEDs treatment of patients with epilepsy may indicate a disturbance of the transsulfuration process of Hcy to Cys in these patients, and diminished transformation of Cys to GSH. Cultured human hepatocytes have demonstrated a transformation of 50% of Cys into GSH [47].

Vitamin B6 is involved in the transsulfuration process of Hcy to Cys, whereas vitamins B2, B12 and FA act as cofactors for its remethylation. A decrease in enzyme or cofactor efficiency, e.g. vitamin B6, may lead to an increase in Cys levels. The study by Apeland et al.. [9] in epileptic patients before B-vitamin supplementation showed an increased total Cys level (total Cys: 260.30±38.62), which was similar to the level in patients with epilepsy in the present study before commencing AEDs therapy. Furthermore, following treatment with B-vitamins, the Cys levels of Apeland et al..'s [9] patients decreased (total Cys: 225.23±41.14) to the level seen in epileptic patients with AEDs treatment in the present study. It seems that B-vitamin supplementation normalized Cys levels and improved the redox status of biothiol.

Homocysteine is redox-active, and its toxic effects have been shown to attribute to direct or indirect perturbation of redox homeostasis [48]. Cysteine is less redox-active than Hcy, and it is converted more slowly and not so completely to its oxidized form and has no demonstrable effect on the glutathione peroxidase activity [48,49]. Upchurch et al.. [49] indicated that cells treated with Hcy show a significant reduction in glutathione peroxidase and this pathway may lead to selective impairment of endothelial cells by this thiol. It is known that, in epileptic patients treated with AEDs, Hcy may lead to endothelium dysfunction [7], evaluation of serum lipids and carotid artery intima media thickness [14,50,51], acceleration of atherosclerosis [8,42], to early stroke or thrombotic events [14], and increase the risk for cardiovascular and cerebrovascular diseases [52].

Perturbations of Hcy and Cys redox-status have been reported after treating pregnant mice intraperitoneally with VPA [53]. In the plasma of these mice, Met and serine levels decreased, while Hcy and Cys concentrations increased. Additionally, a decrease of reduced GSH with no total GSH change has been demonstrated in experimental animals receiving AEDs. The reduced GSH removes hydrogen peroxide in a reaction catalyzed by glutathione peroxidase and the depletion of GSH, a common event in damage-induced apoptosis, is necessary and sufficient to induce cytochrome c release, the key event in apoptotic cell death via this pathway [54]. The results of studies evaluating the effects of VPA on oxidant status in epileptic patients are varied. The study by Yis et al.. [55] in epileptic patients treated with VPA shows an imbalance between oxidant-antioxidant status with unchanged levels of glutathione peroxidase activity. A generation of oxidative stress by VPA as a potentially hepatotoxic factor was demonstrated in the study by Chang & Abbott [56]. On the other hand, Sobaniec et al.. [57] found lower levels of glutathione peroxidase in patients receiving VPA, and Verrotii at al. [58] found that VPA therapy does not appear to cause oxidative stress in epileptic children.

It seems that epileptic patients with active epilepsy can show an increase in cell contents of reactive oxygen species and a low antioxidant status. In patients with progressive myoclonus epilepsy of the Unverrricht-Lundborg type, oxidative stress has been shown to be due to increased intracellular GSH concentration [59]. The present study demonstrated increased levels of GSH both in patients with epilepsy before AEDs treatment with unchanged Cys levels and after treatment with AEDs with decreased Cys level (in particular

CBZ and long-term therapy), which may confirm that oxidative stress participates in the pathogenesis of epilepsy but appears in different mechanisms involving therapy. Generation of oxidative stress in patients with epilepsy before AEDs treatment is also confirmed by the study of Yuksel et al.. [60], which showed increased lipid peroxidation in newly diagnosed patients.

Many studies have demonstrated the success of B-vitamin and FA supplementation in decreasing Hcy levels in epileptic patients treated with AEDs [1,2,34]. However, the study by Sydow et al.. [61], conducted on HHcy patients with peripheral vascular disease, demonstrated that B-vitamin supplementation helps regulate only Hcy levels and doesn't have a sufficient effect in decreasing different risk factors of vascular disease e.g. asymmetric dimethylarginine (ADMA) levels [10]. Moreover, the studies by Sydow et al.. [61] and Koga et al.. [62] indicate that, to regulate both Hcy and ADMA, patients require combined therapy with B-vitamins, FA, and arginine (Arg). The effectiveness of this combined therapy on epileptic patients treated with AEDs appears to also be indicated by the result analysis obtained for Met and Cys after treatment with this therapy in the present study.

Conclusion

The present study demonstrates that AEDs pharmacotherapy rather than epilepsy leads to increases in Hcy levels in epileptic patients, especially in those treated in polytherapy and long-term. In patients with epilepsy before AEDs therapy, the remethylation process of Hcy (decreased level of Met) appears to be disrupted, and after AEDs treatment, especially after VPA therapy, so is the transsulfuration process of this Hcy (resulting in a decreased level of Cys). CBZ monotherapy and AEDs polytherapy and long-term therapy appear to be related to disturbances in both Met:Hcy and Cys:Hcy ratios. Our study indicated that oxidative stress (measured as GSH levels) increased both in patients with epilepsy before AEDs treatment and after treatment with AEDs, in particular CBZ and long-term therapy, but this oxidative stress appears in different mechanisms.

It also seems that, in epileptic patients treated with AEDs, only therapy with B-vitamins and FA can effectively help regulate the levels of the vascular disease factor Hcy and the metabolism of biothiols (Hcy, Met, Cys, GSH).

References

[1] Apeland, T., Mansoor, M. A., Pentieva, K., McNulty, H., Seljeflot, I. & Standjord, R. E. (2002). The effect of B-vitamins on hyperhomocysteinemia in patients on antiepileptic drugs. *Epilepsy Research*, 51, 237-247.

[2] Huemer, M., Ausserer, B., Graninger, G., Hubmann, M., Huemer, C., Schlachter, K., Tscharre, A., Ulmer, H. & Simma, B. (2005). Hyperhomocysteinemia in children treated with antiepileptic drugs is normalized by folic acid supplementation. *Epilepsia*, 46, 1677-1683.

[3] Verrotti, A., Pascarella, R., Trotta, D., Giuva, T., Morgese, G. & Chiarelli, F. (2000). Hyperhomocysteinemia in children treated with sodium valproate and carbamazepine. *Epilepsy Research*, 41, 253-257.

[4] Karabiber, H., Sonmezgoz, E., Ozerol, E., Yakinci, C., Otlu, B. & Yologlu, S. (2003). Effects of valproate and carbamazepine on serum levels of homocysteine, vitamin B12, and folic acid. *Brain and Development*, 25, 113-115.

[5] Attilakos, A., Papakonstantinou, E., Schulpis, K., Voudris, K., Katsarou, E., Mastroyianni, S. & Garoufi, A. (2006). Early effect of sodium valproate and carbamazepine monotherapy on homocysteine metabolism in children with epilepsy. *Epilepsy Research*, 71, 229-232.

[6] Schwaninger, M., Ringleb, P., Winter, R., Kohl, B., Fiehn, W., Rieser, P. A. & Walter-Sack, I. (1999). Elevated plasma concentrations of homocysteine in antiepileptic drug treatment. *Epilepsia*, 40, 345-350.

[7] Apeland, T., Mansoor, M. A. & Strandjord, R. E. (2001). Antiepileptic drugs as independent predictors of plasma total homocysteine levels. *Epilepsy Research*, 47, 27-35.

[8] Hamed, S. A., Hamed, E. A., Hamdy, R. & Nabeshima, T. (2007). Vascular risk factors and oxidative stress as independent predictors of asymptomatic atherosclerosis in adult patients with epilepsy. *Epilepsy Research,* 74, 183-192.

[9] Apeland, T., Froyland, E. S., Kristensen, O., Strandjord, R. E. & Mansoor, M. A. (2008). Drug-induced perturbation of the aminothiol redox-status in patients with epilepsy: improvement by B-vitamins. *Epilepsy Research*, 82, 1-6.

[10] Sniezawska, A., Dorszewska, J., Rozycka, A., Przedpelska-Ober, E., Lianeri, M., Jagodzinski, P.P. & Kozubski, W. (2011). MTHFR, MTR, and MTHFD1 gene polymorphisms compared to homocysteine and asymmetric dimethylarginine concentrations and their metabolites in epileptic patients treated with antiepileptic drugs. *Seizure*, 20, 533-540.

[11] Badiou, S., Breton, H., Peyriere, H., Charasson, V., Crespel, A., Gelisse, P., Cristol, J. P. & Hillaire-Buys, D. (2008). Comparison of carbamazepine and oxcarbamazepine effects on aminothiol levels. *European Journal* of *Clinical Pharmacology,* 64, 83-87.

[12] Gidal, B., Tamura, T., Hammer, A. & Vuong, A. (2005). Blood homocysteine, folate and vitamin B-12 concentrations in patients with epilepsy receiving lamotrigine or sodium valproate for initial monotherapy. *Epilepsy Research*, 64, 161-166.

[13] Yoo, J. H. & Hong, S. B. (1999). A common mutation in the methylenetetrahydrofolate reductase gene is a determinant of hyperhomocysteinemia in epileptic patients receiving anticonvulsants. *Metabolism*, 48, 1047-1051.

[14] Unal, O., Deda, G., Teber, S., Ertem, M. & Akar, N. (2009). Thrombophilic risk factors in epileptic children treated with valproic acid. *Pediatric Neurology*, 40, 102-106.

[15] Sener, U., Zorlu, Y., Karaguzel, O., Ozdamar, O., Coker, I. & Topbas, M. (2006). Effects of common anti-epileptic drug monotherapy on serum levels of homocysteine, vitamin B12, folic acid and vitamin B6. *Seizure*, 15, 79-85.

[16] Caccamo, D., Condello, S., Gorgone, G., Crisafulli, G., Belcastro, V., Gennaro, S., Striano, P., Pisani, F. & Ientile, R. (2004). Screening for C677T and A1298C MTHFR polymorphisms in patients with epilepsy and risk of hyperhomocysteinemia. *Neuromolecular Medicine*, 6, 117-126.

[17] Belcastro, V., Gaetano, G., Italiano, D., Oteri, G., Caccamo, D., Pisani, L. R., Striano, P., Striano, S., Ientile, R. & Pisani, F. (2007). Antiepileptic drugs and MTHFR polymorphisms influence hyper-homocysteinemia recurrence in epileptic patients. *Epilepsia*, 48, 1990-1994.

[18] Mudd, S. H., Skovby, F., Levy, H. L., Pettigrew, K. D., Wilcken, B., Pyeritz, R. E., Andria, G., Boers, G. H., Bromberg, I. L. & Cerone, R. (1985). The natural history of homocystinuria due to cystathionine beta-synthase deficiency. *American Journal of Human Genetics*, 37, 1-31.

[19] Dean, J. C., Robertson, Z., Reid, V., Wang, Q., Hailey, H., Moore, S., Rasalam, A. D., Turnpenny, P., Lloyd, D., Shaw, D. & Little, J. (2008). A high frequency of the MTHFR 677C>T polymorphism in Scottish women with epilepsy: possible role in pathogenesis. *Seizure*, 17, 269-275.

[20] Billings, R. E. (1984). Decreased hepatic 5,10-methylenetetrahydrofolate reductase activity in mice after chronic phenytoin treatment. *Molecular Pharmacology*, 25, 459-466.

[21] Carl, G.F. & Smith, D. B. (1983). The effect of chronic phenytoin treatment on tissue folate concentrations and on the activities of the methyl synthetic enzymes in the rat. *Journal of Nutrition*, 113, 2368-2374.

[22] Hol, F. A., van der Put, N. M., Geurds, M. P., Heil, S. G., Trijbels, F. J., Hamel., B. C., Mariman, E. C. & Blom, H. J. (1998). Molecular genetic analysis of the gene encoding the trifunctional enzyme MTHFD (methylenetetrahydrofolate-dehydrogenase, methenyltetrahydrofolate- cyclohydrolase, formyltetrahydrofolate synthetase) in patients with neural tube defects. *Clinical Genetics*, 53, 119-125.

[23] Dorszewska, J., Florczak, J., Rozycka, A., Kempisty, B., Jaroszewska-Kolecka, J., Chojnacka, K., Trzeciak, W. H. & Kozubski W. (2007). Oxidative DNA damage and level of thiols as related to polymorphisms of MTHFR, MTR, MTHFD1 in Alzheimer's and Parkinson's diseases. *Acta Neurobiologiae Experimentalis*, 67, 113-129.

[24] Ono, H., Sakamoto, A., Mizoguchi, N. & Sakura, N. (2002). The C677T mutation in the methylenetetrahydrofolate reductase gene contributes to hyperhomocysteinemia in patients taking anticonvulsants. *Brain & Development*, 24, 223-226.

[25] Ono, H., Sakamoto, A., Eguchi, T., Fujita, N., Nomura, S., Ueda, H., Sakura, N. & Ueda, K. (1997). Plasma total homocysteine concentrations in epileptic patients taking anticonvulsants. *Metabolism*, 46, 959-962.

[26] Agnati, L.F., Genedani, S., Rasio, G., Galantucci, M., Saltini, S., Filaferro, M., Franco, R., Mora, F., Ferre, S. & Fuxe, K. (2005). Studies on homocysteine plasma levels in Alzheimer's patients. Relevance for neurodegeneration. *Journal of Neural Transmission*, 112, 163-169.

[27] Kubova, H., Folbergrova, J. & Mares, P. (1995). Seizures induced by homocysteine in rats during ontogenesis. *Epilepsia*, 36, 750-756.

[28] Folbergrová, J. (1997). Anticonvulsant action of both NMDA and non-NMDA receptor antagonists against seizures induced by homocysteine in immature rats. *Experimental Neurology*, 145, 442-450.

[29] Folbergrová, J., Druga, R., Otáhal, J., Haugvicová, R., Mares, P. & Kubová, H. (2005). Seizures induced in immature rats by homocysteic acid and the associated brain damage are prevented by group II metabotropic glutamate receptor agonist (2R,4R)-4-aminopyrrolidine-2,4-dicarboxylate. *Experimental Neurology*, 192, 420-436.

[30] Rasić-Marković, A., Stanojlović, O., Hrncić, D., Krstić, D., Colović, M., Susić, V., Radosavljević, T. & Djuric, D. (2009) The activity of erythrocyte and brain Na+/K+ and Mg2+-ATPases in rats subjected to acute homocysteine and homocysteine thiolactone administration. *Molecular* & Cellular *Biochemistry,* 327, 39-45.

[31] Mudd, S. H., Levy, H. L. & Skovby F. (1995). Disorders of transsulfuration. In. C. R. Scriver, A. L. Beaudet, W. S. Sly, & D. Valle (Eds.), The metabolic and molecular bases of inherited disease, (7th edition, pp. 1279-1327), New York, McGraw-Hill.

[32] Tamura, T., Aiso, K., Johnston, K. E., Black, L. & Faught, E. (2000). Homocysteine, folate, vitamin B-12 and vitamin B-6 in patients receiving antiepileptic drug monotherapy. *Epilepsy Research*, 40, 7-15.

[33] Elliott, J. O., Jacobson, M. P. & Haneef. Z. (2007). Homocysteine and bone loss in epilepsy. *Seizure* 16, 22-34.

[34] Bochyńska, A., Lipczyńska-Łojkowska, W., Gugała-Iwaniuk, M., Lechowicz, W., Restel, M., Graban, A., Lipska, B. & Ryglewicz, D. (2012). The effect of vitamin B supplementation on homocysteine metabolism and clinical state of patients with chronic epilepsy treated with carbamazepine and valproic acid. *Seizure*, 21, 276-281.

[35] Centers for Disease Control and Prevention. Recommendations for the use of folic acid to reductase the number of cases of spina bifida and other neural tube defects. (1999). *MMWR* 41(RR-14), 1-7.

[36] Di Giuseppe, D., Di Simplicio, P., Capecchi, P. L., Lazzerini, P. E. & Pasini, F. L. (2003). Alteration in the redox state of plasma in heart-transplant patients with moderate hyperhomocysteinemia. *Journal of Laboratory and Clinical Medicine,* 142, 21-28.

[37] Commision on Classification and Terminology of The International League Against Epilepsy. (1989). Proposal for Revised Classification of Epilepsies and Epileptic Syndromem. *Epilepsia*, 30, 389-399.

[38] Accinni, R., Bartesaghi, S., De Leo, G., Cursano, C. F., Achilli, G., Loaldi, A., Cellerino, C. & Parodi, O. (2000). Screening of homocysteine from newborn blood spots by high-performance liquid chromatography with coulometric array detection. *Journal of Chromatography* A, 896, 186-189.

[39] Moore, J.L. (2005). The significance of folic acid for epilepsy patients. *Epilepsy & Behavior*, 7, 172-181.

[40] Siniscalchi, A., Mancuso, F., Gallelli, L., Ferreri Ibbadu, G., Biagio Mercuri, N. & De Sarro, G. (2005). Increase in plasma homocysteine levels induced by drug treatments in neurologic patients. *Pharmacological Research*, 52, 367-375.

[41] Aslan, K., Bozdemir, H., Unsal, C. & Güvenc, B. (2008). The effect of antiepileptic drugs on vitamin B12 metabolism. *International Journal of Laboratory Hematology*, 30, 26-35.

[42] Tan, T. Y., Lu, C. H., Chuang, H. Y., Lin, T. K., Liou, C. W., Chang, W. N. & Chuang, Y. C. (2009). Long-term antiepileptic drug therapy contributes to the acceleration of atherosclerosis. *Epilepsia*, 50, 1579-1586.

[43] Toohey, J.I. (2001).Possible involvement of sulfane sulfur in homocysteine-induced atherosclerosis. *Medical Hypotheses*, 56, 259-261.

[44] Borowczyk, K., Tisończyk, J. & Jakubowski H. (2012). Metabolism and neurotoxicity of homocysteine thiolactone in mice: protective role of bleomycin hydrolase. *Amino Acids*, 43, 1339-1348.

[45] Bugaj, R., Urbańska, N., Dorszewska, J., Sniezawska, A., Przedpelska-Ober, E. & Kozubski, W. (2011). Homocysteine, cysteine and gluthatione in epileptic patients treated with antiepileptic drugs. *Acta Neurobiologiae Experimentalis*, 71, 1, 149.

[46] Deneke, S. M. (2000). Thiol-based antioxidants. *Current Topics in Cellular Regulation,* 36, 151-180.

[47] Mosharov, E., Cranford, M. R. & Banerjee, R. (2000). The quantitatively important relationship between homocysteine metabolism and glutathione synthesis by the transsulfuration pathway and its regulation by redox changes. *Biochemistry*, 39, 13005-13011.

[48] Zou, C. G. & Banerjee, R. (2005). Homocysteine and redox signaling. *Antioxid Redox Signaling*, 7, 547-559.

[49] Upchurch, G. R. Jr., Welch, G. N., Fabian, A. J., Freedman, J. E., Johnson, J. L., Keaney, J. F. Jr. & Loscalzo, J. (1997). Homocyst(e)ine decreases bioavailable nitric oxide by a mechanism involving glutathione peroxidase. *Journal of Biological Chemistry*, 272, 17012-17017.

[50] Belcastro, V. & Striano, P. (2009). Commentary to: "Evaluation of serum lipids and carotid artery intima media thickness in epileptic children treated with valproic acid". *Brain & Development*, 31, 474.

[51] Ozdemir, O., Yakut, A., Dinleyici, E. C., Aydogdu, S. D., Yarar, C. & Colak, O. (2011). Serum asymmetric dimethylarginine (ADMA), homocysteine, vitamin B(12), folate levels, and lipid profiles in epileptic children treated with valproic acid. *European Journal of Pediatrics*, 170, 873-877.

[52] Mintzer, S., Skidmore, C. T., Abidin, C. J., Morales, M. C., Chervoneva, I., Capuzzi, D. M. & Sperling, M. R. (2009). Effects of antiepileptic drugs on lipids, homocysteine, and C-reactive protein. *Annals of Neurology*, 65, 448-456.

[53] Hishida, R. & Nau, H. (1998). VPA-induced neural tube defects in mice. I. Altered metabolism of sulfur amino acids and glutathione. *Teratogenesis, Carcinogenesis, and Mutagenesis,* 18, 49-61.

[54] Coppola, S. & Ghibelli, L. (2000). GSH extrusion and the mitochondrial pathway of apoptotic signalling. *Biochemical Society Transactions*, 28, 56-61.

[55] Yiş, U., Seçkin, E., Kurul, S. H., Kuralay, F. & Dirik, E. (2009). Effects of epilepsy and valproic acid on oxidant status in children with idiopathic epilepsy. *Epilepsy Research*, 84, 232-237.

[56] Chang, T. K. & Abbott, F. S. (2006). Oxidative stress as a mechanism of valproic acid-associated hepatotoxicity. *Drug Metabolism Reviews,* 38, 627-639.

[57] Sobaniec, W., Solowiej, E., Kulak, W., Bockowski, L., Smigielska-Kuzia, J. & Artemowicz, B. (2006). Evaluation of the influence of antiepileptic therapy on antioxidant enzyme activity and lipid peroxidation in erythrocytes of children with epilepsy. *Journal* of *Child Neurology*, 21, 558-562.

[58] Verrotti, A., Scardapane, A., Franzoni, E., Manco, R., Chiarelli, F. (2008). Increased oxidative stress in epileptic children treated with valproic acid. *Epilepsy Research*, 78, 171-177.

[59] Edwards, M.J., Hargreaves, I.P., Heales, S.J., Jones, S.J., Ramachandran, V., Bhatia, K.P. & Sisodiya, S. (2002). N-acetylcysteine and Unverricht-Lundborg disease: variable response and possible side effects. *Neurology*, 59, 1447-1449.

[60] Yüksel, A., Cengiz, M., Seven, M. & Ulutin, T. (2001). Changes in the antioxidant system in epileptic children receiving antiepileptic drugs: two-year prospective studies. *Journal* of *Child Neurology,* 16, 603-606.

[61] Sydow, K., Schwedhelm, E., Arakawa, N., Bode-Boger, S.M., Tsikas, D., Hornig, B., Frolich, J.C. & Boger, R.H. (2003). ADMA and oxidative stress are responsible for endothelial dysfunction in hyperhomocyst(e)inemia: effects of L-arginine and B vitamins. *Cardiovascular Research*, 57, 244-252.

[62] Koga, Y., Akita, Y., Nishioka, J., Yatsuga, S., Povalko, N., Tanabe, Y., Fujimoto, S. & Matsuishi, T. (2005). L-Arginine improves the symptoms of strokelike episodes in Melas. *Neurology*, 64, 710-712.

Index

A

B

C

D

G

H

M

N

O

P

Q

R

S

T

U

V

W

X

Y